# Waste Management as Economic Industry Towards Circular Economy

Sadhan Kumar Ghosh
Editor

# Waste Management as Economic Industry Towards Circular Economy

 Springer

*Editor*
Sadhan Kumar Ghosh
Department of Mechanical Engineering
Jadavpur University
Kolkata, West Bengal, India

ISBN 978-981-15-1619-1          ISBN 978-981-15-1620-7    (eBook)
https://doi.org/10.1007/978-981-15-1620-7

This Springer imprint is published by the registered company Springer Nature Singapore Pte Ltd.
The registered company address is: 152 Beach Road, #21-01/04 Gateway East, Singapore 189721, Singapore

# Preface

Japan's 2020 Olympic and Paralympic teams to wear sportswear made from recycled clothes supporting the circularity of resources while the linear economic model "take–make–dispose" is reaching supply and demand limitations, posing risks in many areas of interventions. A circular mindset allows companies to move beyond mitigating these risks to capture the USD$ 4.5 trillion economic opportunities that the circular economy represents. In the period from 1970 to 2017, the global extraction of biotic and abiotic materials increased more than 240%, by 65 billion tons since 1970, reaching an amount of 92 billion tons in 2017 as per the IPCC data. Continued extraction of finite resources, climate changes and demographic changes will make it harder for businesses to provide the products and services necessary for a well-functioning global economy. Successful circular initiatives will reduce the dependence on dwindling natural resources and will create value for companies and their stakeholders.

The results of interviews and survey responses of more than 100 CEOs of different industries and other organizations conducted by World Business Council for Sustainable Development (WBCSD) identified eight business cases to justify circular economy practices: 1) create additional revenue from existing products and processes, 2) spur innovation of new products and services, 3) reduce operating costs, 4) engage customers and employees, 5) distinguish from competition, 6) align with corporate strategy or mission, 7) adapt business models and value chain relationships and 8) mitigate linear risk exposure. Business cases for moving to the circular economy can be classified under the following categories: driving growth, enhancing competitiveness and mitigating risk.

Raw materials are used in building and processing in the factories, transforming them into new products and consumer goods and at the end of the lifetime of their use disposed to landfills. The International Society of Waste Management, Air and Water (ISWMAW) and the International Solid Waste Association (ISWA) recognize that in low-income countries the informal and micro-enterprise recycling, reuse and repair systems achieve significant recycling rates of 20–30%. This activity saves local authorities around 20% or more of what they would otherwise spend additionally on waste management, representing many millions of dollars every year in large cities. All these types of waste are being considered for recycling to compensate for the loss in resources reserve. The potential recovery of rare earths from scrap electronics could become a substantial source of supply, with real business opportunities for companies that look to reuse these metals, rather than paying the high prices demanded for virgin materials. The wide range of materials consumed can be aggregated into four main categories: biomass, minerals, fossil fuels and metals. In 2015, around 84.4 Gt of raw materials such as minerals, biomass or fossil fuels were extracted worldwide, and only 8.4 Gt (about 10%) of recycled materials re-entered the economic system. This relationship underlines the considerable potential of circularity to shift from extraction to reuse and recycling of raw materials in all the spheres of materials used.

Given the high volatility of resource prices, heavy pollution of primary production and above all the resource productivity and sustainable use of natural resources, recycling becomes mandatory. Waste management in the society, industries, commercial and household establishments and municipalities will be effective if the waste is treated as a resource considering a business model following the circularity of resources instead of the linear economic model of "take–make–dispose". The overall goals are resource conservation and environmental protection, as well as generating economic benefits. Material resources remain a major pillar in economic activities. As per UN Global Compact (UNGC, India), more than 50% of the progress towards the SDGs will come from India. In parallel, India presents 25% of the $4 trillion worth of market opportunities for companies working in the sustainable area globally and employment generation potential of 72 million by 2030. In the context of Agenda 2030, we see a striking commitment to sustainable development. Ninety-one percentage of Indian CEOs (87% globally) believe the SDGs provide an essential opportunity for business to rethink approaches to sustainable value creation.

The 8th IconSWM 2018 has received 380 abstracts and 320 full papers from 30 countries. Three hundred accepted full papers have been presented as oral and poster presentations in November at ANU, Guntur, AP, India. The chapters finally selected by the board have been thoroughly reviewed by the experts for the book "Waste Management as Economic Industry Towards Circular Economy" dealing with the women empowerment, city-based waste management, benefit of CE in South Africa, sustainable product development in manufacturing, biodegradable packaging materials, wealth from poultry waste, waste plastics to useful products, upcycling of scraps, business model for ERP by establishing a sustainable supply chain management, economic implications of waste recycling, secondary raw materials from waste, sustainable waste transportation, waste management through public–private partnership project, opportunities for resource recovery of biodegradable MSW, wastewater treatment, resource efficiency and CE, green temples—CE in waste management, faecal sludge treatment, CE, etc.

The IconSWM movement was initiated for better waste management, resource circulation and environmental protection since the year 2009 through generating awareness and bringing all the stakeholders together from all over the world under the aegis of the International Society of Waste Management, Air and Water

(ISWMAW). It established a few research projects across the country those include CST at the Indian Institute of Science, Jadavpur University, KIIT, Calcutta University, etc. The Consortium of Researchers in International Collaboration (CRIC) and many other organizations across the world help in the IconSWM-CE movement. IconSWM has become significantly one of the biggest platforms in India for knowledge sharing, awareness generation and encouraging the urban local bodies, government departments, researchers, industries, NGOs, communities and other stakeholders in the area of waste management. The primary agenda of this conference is to reduce the waste generation encouraging the implementation of Reduce, Reuse and Recycle (3Rs) and circular economy concept and management of the generated waste ensuring resource circulation. The conference will show a paradigm and provide holistic pathways to waste management and resource circulation conforming to urban mining and circular economy.

The success of the 8th IconSWM is the result of significant contribution of many organizations and individuals, specifically the Government of Andhra Pradesh, several industry associations and chamber of commerce and industries, the AP State Council of Higher Education, Swachh Andhra Mission and various organizations in India and in different countries as our partners including UNEP, UNIDO and UNCRD. The 8th IconSWM 2018 was attended by nearly 823 delegates from 22 countries. The 9th IconSWM-CE 2019 was held at KIIT, Bhubaneswar, Odisha, during 27–30 November 2019 participated by 21 countries. Sri Venkateswara University (SVU) has expressed their willingness to organize the 10th IconSWM-CE at SVU in the temple city Tirupati, Andhra Pradesh, tentatively during 2–5 December 2020. This book will be helpful for the researchers, educational and research institutes, policymakers, government, implementers, ULBs and NGOs. Hope to see you all in the 10th IconSWM-CE 2020 at Tirupati.

Kolkata, India                                              Prof. Sadhan Kumar Ghosh
February 2020

# Acknowledgements

I thank Hon'ble Chief Minister and Hon'ble Minister of MA&UD, Government of Andhra Pradesh, for taking personal interest in this conference.

I am indebted to Shri. R. Valavan Karikal, IAS; Dr. C. L. Venkata Rao; Shri. B. S. S. Prasad, IFS (Retd.); Prof. S. Vijaya Raju; and Prof. A. Rajendra Prasad, VC, ANU, for their unconditional support and guidance for preparing the platform for the successful 8th IconSWM at Guntur, Vijayawada, AP. The support of UNCRD/DESA is being gratefully acknowledged as the 8th IconSWM-CE has been declared as the official pre-event of the 9th Regional 3R Forum in the Asia and the Pacific held during 4–6 March 2019 at Bangkok, Thailand.

I must express my gratitude to Mr Vinod Kumar Jindal, ICoAS; Shri. D. Muralidhar Reddy, IAS; Shri, K. Kanna Babu, IAS; Mr. Vivek Jadav, IAS; Mr. Anjum Parwez, IAS; Prof. S. Varadarajan, Mr. Bala Kishore and Mr. K. Vinayakam; Prof. Shinichi Sakai, Kyoto University, JSMCWM; Prof Y. C. Seo and Prof S. W. Rhee, KSWM; Shri C R C Mohanty, UNCRD; members of Industry Associations in Andhra Pradesh; Prof. P. Agamuthu, WM&R; Prof. M Nelles, University of Rostock; Dr. Rene Van Berkel, UNIDO; Ms. Kakuko Nagatani-Yoshida and Mr. Atul Bagai, UNEP and UN Delegation to India, for their active support.

The IconSWM-ISWMAW Committee acknowledges the contribution and interest of all the sponsors, industry partners, industries, co-organizers, organizing partners around the world, the Government of Andhra Pradesh, Swachh Andhra Corporation as the principal collaborator, vice chancellor and all the professors and academic community at Acharaya Nagarjuna University (ANU), chairman, vice chairman, secretary and other officers of AP State Council of Higher Education for involving all the universities in the state, chairman, member secretary and officers of the AP Pollution Control Board, director of factories, director of boilers, director of mines and officers of different ports in Andhra Pradesh and the delegates and service providers for making a successful 8th IconSWM.

I must specially mention the support and guidance by each of the members of the International Scientific Committee, CRIC members, the core group members and the local organizing committee members of the 8th IconSWM who are the pillars for the success of the programme. The editorial board members including the reviewers, authors and speakers and Mr Aninda Bose and the team of M/s. Springer India Pvt. Ltd deserve thanks, who were very enthusiastic in giving me inputs to bring this book.

I must mention the active participation of all the team members in IconSWM secretariat across the country with special mention of Prof. H.N. Chanakya and his team in IISc Bangalore; Ms. Sheetal Singh and Dr. Sandhya Jaykumar and their team in CMAK and BBMP; Mr. Saikesh Paruchuri, Mr. Anjaneyulu, Ms. Senophiah Mary, Mr Rahul Baidya, Dr. Asit Aich, Ms. Ipsita Saha, Mr. Suresh Mondal, Mr. Bisweswar Ghosh, Mr. Gobinda Debnath and the research team members in Mechanical Engineering Department and ISWMAW, Kolkata HQ for various activities for the success of the 8th IconSWM 2018. I express my special thanks to Sannidhya Kumar Ghosh, being the governing body member of ISWMAW supported the activities from USA. I am indebted to Mrs Pranati Ghosh who gave me guidance and moral support in achieving the success of the event. Once again, IconSWM and ISWMAW express gratitude to all the stakeholders, delegates and speakers who are the part of the success of the 8th IconSWM 2018.

# Contents

# About the Editor

**Dr. Sadhan Kumar Ghosh** is the Dean, Faculty of Engineering and Technology, Professor & Former Head of the Mechanical Engineering Department and Founder Coordinator of the Centre for QMS at Jadavpur University, India. A prominent figure in the fields of waste management, circular economy, SME sustainability, green manufacturing, green factories and TQM, he served as the Director, CBWE, Ministry of Labour and Employment, Government of India, and L&T Ltd. Professor Ghosh is also the Founder and Chairman of the IconSWM and President of the International Society of Waste Management, Air and Water, as well as the Chairman of the Indian Congress on Quality, Environment, Energy and Safety Management Systems (ICQESMS).

He was awarded a Distinguished Visiting Fellowship by the Royal Academy of Engineering, UK, to work on "Energy Recovery from Municipal Solid Waste" in 2012. He received the Boston Pledge and NABC 2006 Award for the most eco-friendly innovation "Conversion of plastics & jute waste to wealth" in the EssSP/50K Business Plan Competition at Houston, Texas, USA. He holds patents on waste plastic processing technology and high-speed jute ribboning technology preventing water wastage and occupational hazards. He is member of ISO Working Groups concerning waste management (ISO/TC 297). His projects have been funded by European Union Horizon 2020 (2018–2022) on waste water, Royal Academy of Engineering (2018–2020 & 2012) on Circular economy in SMEs, Shota Rustaveli National Science Foundation (SRNSF) of Georgia (2019–2021) on resource circulation from landfill, GCRF 2019/2020 Pump Priming – Aston Project UK on impact of wellbeing & mental health on productivity & sustainability in Industries, UNCRD/DESA as Expert (2016–2018) on SWM, Asian Productivity Organisation (APO) (2016–2019) on green manufacturing, British Council & DST (2012–2014), Royal Society (2015), Erasmus Plus (2016–2017), ISWMAW (2018–2021), Indian Statistical Institute (2019–2021), Institute of Global Environmental Strategies (IGES, Japan) (2019), South Asian Cooperation Environmental

Programme (SACEP, Sri Lanka)(2018–2020) for preparing SWM roadmap for South Asian countries, Jute Technology Mission (2008–2011), Central Pollution Control Board (1999–2002), Government of India on plastic waste management and a few others.

# Circular Economy Approach to Women Empowerment Through Reusing Treated Rural Wastewater Using Constructed Wetlands

**B. Lekshmi, Shruti Sharma, Rahul S. Sutar, Yogen J. Parikh, Dilip R. Ranade and Shyam R. Asolekar**

**Abstract** Integrating environmental activity and economic development is one of the key milestones in the circular economy. There is an urgent need for developing countries like India to step path towards a circular economy for its sustainable development and environmental improvement. Water industry in India, there should move towards a circular economy model from the linear economy model. To functionalize circular economy through rural development, a new sustainable business model has been proposed in this study by employing natural treatment systems for wastewater treatment. Natural treatment systems especially constructed wetlands for wastewater treatment and reuse are the eco-friendliest technologies currently practised worldwide. In India, there is a lack of successful demonstrations in the field of wastewater treatment using constructed wetland and its gainful utilisation. In this present study, a successful circular economy model is demonstrated through the case study in Mhaswad Town which is located in District of Satara, State of Maharashtra and has a population of ~35,000. Wastewater generated (max 250 m$^3$/d) from half Mhaswad population, i.e. ~14,000 is treated using horizontal subsurface flow constructed wetland which is "CW4Reuse" technology developed by IIT Bombay. The treatment plant consists of settling tank followed by nine beds of constructed wetland treatment units. Currently, the treatment plant is meeting the standards for treated water recommended by the pollution control board. The treatment plant is operated by women members of a cooperative society in Mhaswad. It is envisaged that the cooperative society will be trading the treated water for profit and thus they will be earning money through waste management. The model predicts net positive gain by treating wastewater through constructed wetland compared to conventional treatment method. This will facilitate rural development as well as women empowerment by implementing eco-centric treatment technology.

B. Lekshmi (✉) · S. Sharma · R. S. Sutar · Y. J. Parikh · D. R. Ranade · S. R. Asolekar (✉)
Centre for Environmental Science & Engineering, Indian Institute of Technology Bombay, Mumbai, India
e-mail: b.lekshmi.nair@gmail.com

S. R. Asolekar
e-mail: asolekar@gmail.com

**Keywords** Wastewater treatment · Water reuse · Natural treatment systems · Constructed wetlands · Circular economy · Rural development

# 1 Introduction

Environmental glitches in urban or in rural, local or global are mainly due to the linearity of economy of human (Masullo 2017). Circular economy approach will help in development of sustainable, low carbon footprint, more resource efficient systems as well as a competitive economy (EU Commission 2015). In recent years, many countries have introduced circular economy approach for wastewater treatment systems (Masi et al. 2018; Abu-Ghunmi et al. 2016; Eshetu Moges et al. 2018). Wastewater treatment using natural treatment systems like constructed wetlands was reported by various researchers for several applications like domestic wastewater treatment (Kumar and Asolekar 2016; Kumar et al. 2016), removal pesticides from wastewater and reuse of water in agriculture (Vymazal and Březinová 2015), for treatment of agro-industrial wastewaters (Tatoulis et al. 2015), storm water treatment (Adyel et al. 2016), treatment of combined sewer overflows (Henrichs et al. 2007; Masi et al. 2017), treatment of landfill leachate (Mojiri et al. 2016), distillery effluent (Billore et al. 2001) and industrial wastewater treatment (Vymazal 2014).

Due to high operation and maintenance costs of conventional and advanced wastewater treatment technologies, development of alternate technologies in country like India is inevitable. Also, there is a long gap between water demand and supply in urban as well as in rural India. In this context, development of eco-friendly, low-cost and highly efficient treatment technology is adequate for small communities. Natural treatment systems help to fulfil this gap. Therefore, constructed wetlands (CWs) as decentralised system come into picture. This technology is typically classified under the so-called natural treatment systems. Some of the other treatment technologies in this class happen to be waste stabilization ponds, duckweed ponds, oxidation ditches, etc. (Arceivala and Asolekar 2007). Horizontal subsurface flow constructed wetlands (HSSF-CWs) are one of the most common engineered wetlands used in treatment of domestic wastewater. Several worldwide studies have reported that the CWs found to be the efficient technology in treatment of wastewater by removal of organics, nutrients, suspended solids, pathogens, heavy metals including other emerging contaminants. Wastewater treatment using natural treatment systems, especially using the CWs, has been a favoured technology in India.

Present study aims at demonstrating the role of constructed wetlands in wastewater treatment targeting the recycle of water, delivering services to the ecosystem, demonstrating the development of resources from generated wastes and their by empowering the women community. Integrating circular economy concept into the system is the new paradigm demonstrated here in this study. For economic development and improvement in water industry, India ought to upgrade from end of pipe approach, i.e. linear economic model to circular economic models. This study demonstrates a circular economy model in wastewater sector for rural development as well as women empowerment.

## 2  Old Paradigm of Wastewater Management in Mhaswad Town

Direct discharge of wastewaters from industries, households and other wastes into surface waterbodies or rivers are the main causes of degrading water quality of such resources (Kumar et al. 2015a, b; Asolekar and Gopichandran 2005). A case study to demonstrate the treatment of domestic wastewater discharged into river and there by exhibiting the circular economy of wastewater treatment is presented here.

Mhaswad Town in Maan taluka is one of the oldest cities in the State of Maharashtra, India. The town is located in the banks of Manganga River which have 17 wards and a population of around 24,000 inhabitants (Census 2011). Manganga River is a drinking water source for nearby town as well. Projected population for the year 2018 in the town is around 35,000 inhabitants. There were no wastewater management systems previously established in the town. Wastewater generated in the town was previously discharged into the Manganga River and thus causing the pollution of the river basin and thus the drinking water resource. The town was also facing water scarcity and immense drought for around past 25 years. There is high demand of water in the town for non-potable uses such as water municipal garden in Mhaswad Town, small-scale household agricultural activities, livestock farming, etc. There was an urgent need of wastewater management system as well as source of water for the town. The past scenario of water demand and wastewater management in Mhaswad Town is depicted in Fig. 1, and the pictorial representation of flow of wastewater in Mhaswad Town is depicted in Fig. 2.

There is huge potential of resource recovery from wastewaters. The main resources that can be recovered from wastewater include organic matter, nutrients, energy and

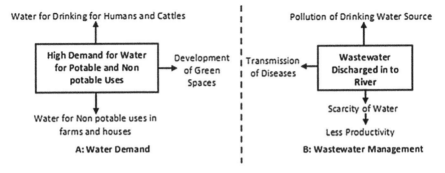

**Fig. 1** Past scenario of water demand and wastewater management in Mhaswad Town

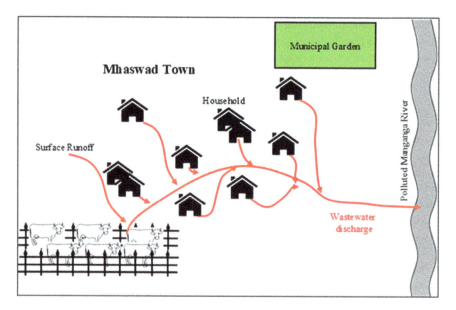

**Fig. 2** Past scenario of flow of wastewater in Mhaswad Town

finally the water itself. But to recognise the most sustainable solution for wastewater treatment is one of the greatest challenges (van der Hoek et al. 2016). In Mhaswad Town as well, the wastewater generated can be used as a resource and the present study demonstrates the successful management of wastewater using eco-friendly, low-cost wastewater treatment technology. The following sections describe the new paradigm of waste management in the town.

## 3  New Paradigm of Wastewater Management in Mhaswad Town

Removal of organic pollutants, nutrients, pathogens, micropollutants, heavy metals, etc. (Vymazal 2010; Zhang et al. 2014; Sartori et al. 2016; Wang et al. 2016; Zhang et al. 2017) from wastewater using different configurations of constructed wetlands in macro-, micro- or mesocosm scale were exclusively studied by various researchers in past and present years (Xinshan et al. 2010; Bustamante et al. 2011; Butterworth et al. 2016; Hu et al. 2016). In horizontal subsurface flow CWs (HSSF-CWs), wastewater is intended to stay behind the media (substrate), flow of the water will be in and around media, root and rhizome of the plants (Kadlec and Wallace 2009). The main mechanism of wastewater treatment using constructed wetland is through many processes like organic matter degradation by microorganism inside the wetland bed, nutrient uptake by the wetland vegetation, pathogen removal by filtration and other physicochemical processes (Vymazal et al. 1998; Vymazal 2007).

CWs are eco-friendly, cost-effective treatment technology for wastewater especially in peri-urban and rural areas in India. However, CWs treatment technology is less explored in India and also there are less successful real-life WWTPs currently operated with this technology for domestic wastewater (Kumar and Asolekar 2016). Horizontal subsurface flow constructed wetland system for treatment of wastewater is developed at IIT Bombay and the developed system is known as "CW4Reuse" technology (Kumar and Asolekar 2016; Kumar et al. 2015a). The components of "CW4Reuse" technology developed are previously described in Kumar and Asolekar (2016).

A portion of wastewater generated ($Q = 18$–$25$ m$^3$/d) from around 14,000 populations from the present study area, Mhaswad Town is collected and treated in the "CW4Reuse" wastewater treatment plant. Wastewater is collected in the sewage sump with a capacity of around 30 m$^3$. The wastewater overflow is directed to the secondary treatment unit, i.e. "CW4Reuse". This wastewater treatment unit consists of nine beds of HSSF-CWs each having the dimensions in the range of 30: 6.7: 1 (L: W: D). The treated wastewater is collected in a treated water collection tank with capacity of around 35 m$^3$. The treatment plant is currently owned by Mann Deshi Foundation operated by the women entrepreneurs of the Foundation.

The wastewater quality of the treatment systems, i.e. inlet and outlet water quality is monitored monthly. The water quality parameters of inlet wastewater are typically in the range of Indian domestic wastewater quality. The outlet water quality parameters of wastewater treated using "CW4Reuse" is meeting the discharge standards prescribe by USEPA standard limits for reuse as non-potable water. The current scenario of wastewater management in Mhaswad Town is depicted in Fig. 3. The treated water from the wastewater treatment plant is currently used in the nearby municipal garden. And intermittently water is pumped from the collection tank and is transported to the end users such as fodder camp facility and households. Finally, any excess water is discharged into the river. The constructed wetland treatment facility is currently serving as a water resource for the town of Mhaswad. The harvested plants from the wetland bed are used as livestock fodder in the town. The future benefits of the wastewater management system employed are further discussed in the following section.

# 4 Constructed Wetlands as Ecosystem Service and Economic Resource

Constructed wetlands for wastewater treatment systems were well accepted in the countries like Europe, North America, Australia, etc. (Vymazal 2011). It also has huge potential in countries like India where there is huge gap between supply and demand of water (Kumar et al. 2016; Mahmood et al. 2013; Starkl et al. 2013). There is huge potential of application of constructed wetlands especially in rural areas (Kivaisi 2001). Ecological, economic and social benefits of constructed wetlands

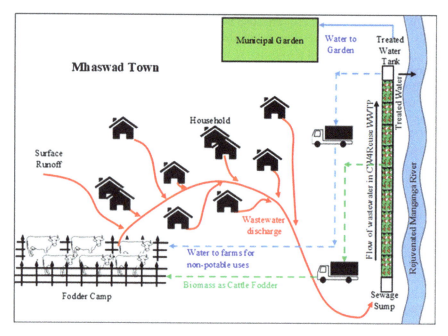

**Fig. 3** Closing the loop: present scenario of wastewater treatment in Mhaswad Town

as a wastewater management system are huge (Masi et al. 2018). The wastewater treatment system "CW4Reuse" in Mhaswad Town is a typical practical example of such a model. The wastewater management system, i.e. "CW4Reuse" technology is an ecological service, social as well as an economic payback in the town. Figure 4

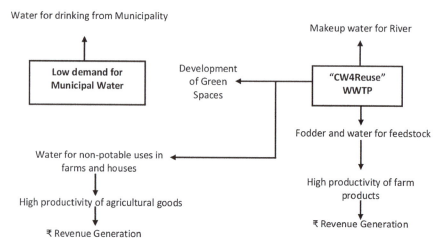

**Fig. 4** Business model of wastewater management in Mhaswad Town

represents the future business model of wastewater management in Mhaswad Town.

The ecological benefits of the wastewater treatment strategy using constructed wetlands include rejuvenation of the Manganga River, which is a source of water for downstream towns. The CW employed is preventing the major source of pollution in the river. The constructed wetlands also contribute to the improvement of microclimate and serve as new habitat for flora and fauna. Natural treatment systems like constructed wetlands are having its own distinct ecosystem that support propagation of wild habitat and help to improve recreational facilities (Starkl et al. 2013). The horizontal subsurface flow constructed wetlands have additional benefits such as no mosquito breeding.

The CW4Reuse treatment plant in the Mhaswad Town is currently owned by women entrepreneurs in the town and it is envisaged that the workers will trade the treated water for profit. This will be one of the sources of their income. Similarly, there will be more water supply available for cattle livestock breeding and there by higher productivity. This will empower the women community as the treatment plant will be completely owned by them. There are no operating costs for the treatment plant. The operating costs addressed are mainly for transporting water to the end users. It is also envisaged that, when there is more supply of water and fodder for livestock production there will be more job opportunities in Mhaswad Town. Social, economic and ecological benefits of constructed wetlands as a wastewater treatment technology are depicted in Fig. 5.

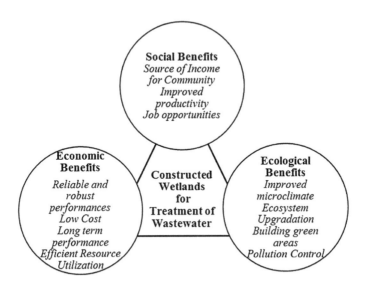

**Fig. 5** Social, economic and ecological benefits of constructed wetlands

# 5 Conclusions and Recommendations

Low-cost treatment technologies like natural treatment systems can address issues of pollution due to wastewater discharges and their applications are diverse. Various researchers have proved the application of CWs in treatment of effluents from different origins such as domestic, industrial, agricultural and hospital. Present study successfully demonstrates the application of constructed wetlands in rural areas. Constructed wetlands as a secondary treatment technology for wastewater should be explored more in Indian context. In recent years, research is focused in the areas of micropollutant removal using constructed wetlands. More research will be done in the future for the present study in that direction. It is recommended that low-cost natural treatment systems should be implemented for sustainable development of the country at large.

**Acknowledgements** Authors of this paper gratefully acknowledge the funding obtained from the two agencies for this research, namely: Indian Institute of Technology Bombay (IITB) and Rajiv Gandhi Science and Technology Commission (RGSTC).

# References

Abu-Ghunmi, D., Abu-Ghunmi, L., Kayal, B., & Bino, A. (2016). Circular economy and the opportunity cost of not 'closing the loop' of water industry: The case of Jordan. *Journal of Cleaner Production, 131,* 228–236.

Adyel, T. M., Oldham, C. E., & Hipsey, M. R. (2016). Stormwater nutrient attenuation in a constructed wetland with alternating surface and subsurface flow pathways: Event to annual dynamics. *Water Research, 107,* 66–82.

Arceivala, S. J., & Asolekar, S. R. (2007). *Wastewater treatment for pollution control and reuse* (3rd ed., 11th Reprint). New Delhi: McGraw Hill Education India Pvt. Ltd.

Asolekar, S. R., & Gopichandran, R. (2005). *Preventive environmental management—An Indian Perspective.* New Delhi: Foundation Books Pvt. Ltd. (The Indian associate of Cambridge University Press, UK).

Billore, S. K., Singh, N., Ram, H. K., Sharma, J. K., Singh, V. P., Nelson, R. M., et al. (2001). Treatment of a molasses based distillery effluent in a constructed wetland in central India. *Water Science and Technology, 44*(11–12), 441.

Bustamante, M. A. O., Mier, M. V., Estrada, J. A. E., & Domíguez, C. D. (2011). Nitrogen and potassium variation on contaminant removal for a vertical subsurface flow lab scale constructed wetland. *Bioresource Technology, 102*(17), 7745–7754.

Butterworth, E., Richards, A., Jones, M., Mansi, G., Ranieri, E., Dotro, G., et al. (2016). Performance of four full-scale artificially aerated horizontal flow constructed wetlands for domestic wastewater treatment. *Water, 8*(9), 365.

Eshetu Moges, M., Todt, D., & Heistad, A. (2018). Treatment of source-separated blackwater: A decentralized strategy for nutrient recovery towards a circular economy. *Water, 10*(4), 463.

EU Commission. (2015). Communication from the Commission to the European Parliament, The Council, The European Economic and Social Committee and the Committee of the Regions, Closing the Loop—An EU Action Plan for the Circular Economy.

Henrichs, M., Langergraber, G., & Uhl, M. (2007). Modelling of organic matter degradation in constructed wetlands for treatment of combined sewer overflow. *Science of the Total Environment, 380*(1–3), 196–209.

Hu, Y., He, F., Ma, L., Zhang, Y., & Wu, Z. (2016). Microbial nitrogen removal pathways in integrated vertical-flow constructed wetland systems. *Bioresource Technology, 207,* 339–345.

Kadlec, R. H., & Wallace, S. (2009). *Treatment wetlands.* Boca Raton: CRC Press.

Kivaisi, A. K. (2001). The potential for constructed wetlands for wastewater treatment and reuse in developing countries: A review. *Ecological Engineering, 16*(4), 545–560.

Kumar, D., & Asolekar, S. R. (2016). Experiences with laboratory and pilot scale constructed wetlands for treatment of domestic wastewaters and effluents. Natural Water Treatment Systems for Safe and Sustainable Water Supply in the Indian Context: Saph Pani, 149.

Kumar, D., Asolekar, S. R., & Sharma, S. K. (2015a). Post-treatment and reuse of secondary effluents using natural treatment systems: The Indian practices. *Environmental Monitoring and Assessment, 187,* 612.

Kumar, D., Chaturvedi, M. K. M., Sharma, S. K., & Asolekar, S. R. (2015b). Sewage-fed aquaculture: A sustainable approach for wastewater treatment and reuse. *Environmental Monitoring and Assessment, 187,* 656.

Kumar, D., Sharma, S. K., & Asolekar, S. R. (2016). Significance of incorporating constructed wetlands to enhance reuse of treated wastewater in India. Natural Water Treatment Systems for Safe and Sustainable Water Supply in the Indian Context: Saph Pani, 161.

Mahmood, Q., Pervez, A., Zeb, B. S., Zaffar, H., Yaqoob, H., Waseem, M., et al. (2013). *Natural treatment systems as sustainable ecotechnologies for the developing countries.* BioMed Research International.

Masi, F., Bresciani, R., Rizzo, A., & Conte, G. (2017). Constructed wetlands for combined sewer overflow treatment: Ecosystem services at Gorla Maggiore, Italy. *Ecological Engineering, 98,* 427–438.

Masi, F., Rizzo, A., & Regelsberger, M. (2018). The role of constructed wetlands in a new circular economy, resource oriented, and ecosystem services paradigm. *Journal of Environmental Management, 216,* 275–284.

Masullo, A. (2017). Organic wastes management in a circular economy approach: Rebuilding the link between urban and rural areas. *Ecological Engineering, 101,* 84–90.

Mojiri, A., Ziyang, L., Tajuddin, R. M., Farraji, H., & Alifar, N. (2016). Co-treatment of landfill leachate and municipal wastewater using the ZELIAC/zeolite constructed wetland system. *Journal of Environmental Management, 166,* 124–130.

Sartori, L., Canobbio, S., Fornaroli, R., Cabrini, R., Marazzi, F., & Mezzanotte, V. (2016). COD, nutrient removal and disinfection efficiency of a combined subsurface and surface flow constructed wetland: A case study. *International Journal of Phytoremediation, 18*(4), 416–422.

Starkl, M., Amerasinghe, P., Essl, L., Jampani, M., Kumar, D., & Asolekar, S. R. (2013). Potential of natural treatment technologies for wastewater management in India. *Journal of Water Sanitation and Hygiene for Development, 3*(4), 500–511.

Tatoulis, T., Stefanakis, A. I., Akratos, C. S., Terkerlekopoulou, A., Gianni, A., Zacharias, I., et al. (2015, September). Treatment of agro-industrial wastewaters using novel horizontal subsurface constructed wetlands. In *6th International Symposium on Wetland Pollutant Dynamics and Control,* York, UK (pp. 13–18).

van der Hoek, J. P., de Fooij, H., & Struker, A. (2016). Wastewater as a resource: Strategies to recover resources from Amsterdam's wastewater. *Resources, Conservation and Recycling, 113,* 53–64.

Vymazal, J. (2007). Removal of nutrients in various types of constructed wetlands. *Science of the Total Environment, 380*(1–3), 48–65.

Vymazal, J. (2010). Constructed wetlands for wastewater treatment. *Water, 2*(3), 530–549.

Vymazal, J. (2011). Constructed wetlands for wastewater treatment: Five decades of experience. *Environmental Science and Technology, 45*(1), 61–69.

Vymazal, J. (2014). Constructed wetlands for treatment of industrial wastewaters: A review. *Ecological Engineering, 73,* 724–751.

Vymazal, J., & Březinová, T. (2015). The use of constructed wetlands for removal of pesticides from agricultural runoff and drainage: A review. *Environment International, 75,* 11–20.

Vymazal, J., Brix, H., Cooper, P. F., Haberl, R., Perfler, R., & Laber, J. (1998). Removal mechanisms and types of constructed wetlands. In: *Constructed wetlands for wastewater treatment in Europe* (pp. 17–66).

Wang, W., Ding, Y., Ullman, J. L., Ambrose, R. F., Wang, Y., Song, X., et al. (2016). Nitrogen removal performance in planted and unplanted horizontal subsurface flow constructed wetlands treating different influent COD/N ratios. *Environmental Science and Pollution Research, 23*(9), 9012–9018.

Xinshan, S., Qin, L., & Denghua, Y. (2010). Nutrient removal by hybrid subsurface flow constructed wetlands for high concentration ammonia nitrogen wastewater. *Procedia Environmental Sciences, 2,* 1461–1468.

Zhang, D. Q., Jinadasa, K. B. S. N., Gersberg, R. M., Liu, Y., Ng, W. J., & Tan, S. K. (2014). Application of constructed wetlands for wastewater treatment in developing countries—A review of recent developments (2000–2013). *Journal of Environmental Management, 141,* 116–131.

Zhang, Y., Lv, T., Carvalho, P. N., Zhang, L., Arias, C. A., Chen, Z., et al. (2017). Ibuprofen and iohexol removal in saturated constructed wetland mesocosms. *Ecological Engineering, 98,* 394–402.

# Waste Management in Balikpapan City Based on Circular Economy

**Suryanto Ibrahim**

**Abstract** Balikpapan is classified as a city with rapid growth and development. The speed of development of the city is influenced by the rapid growth of population. The development of this city and rapid population growth demand the need for adequate land. Conditions like this, on the one hand, benefit the economic growth of Balikpapan city but on the other hand affect the carrying capacity and capacity of the environment. This reality demands the city government to be able to better manage and control various aspects related to the environment as a whole. Various development policies compiled in various city development plans must always be matched with environmental conditions, so that environmental and sustainable development can be realized. Every citizen has the right of a good and healthy environment. If environmental conditions are not good and unhealthy, it will cause various problems. That is why we needed good environmental protection and management. This is consistent with the affirmation of Law Number 32 of 2009 concerning Environmental Protection and Management Article 1 Number 2 which states that: "Protection and management of the environment is an integrated systematic effort carried out to preserve environmental functions and prevent pollution and/or environmental damage, which includes planning, utilization, control, maintenance, supervision and law enforcement. In order to support this, it is necessary to carry out environmental management efforts in a sustainable manner, which can provide maximum benefit for human welfare and not cause environmental damage through policies in environmental management". Balikpapan city government already made various efforts which are supported by the public and the private sectors in this city. Broadly speaking, these efforts are directed to realize the condition of clean land, clean water, clean air and sufficient land cover.

**Keyword** Chain

S. Ibrahim (✉)
Balikpapan Environmental Agency, Balikpapan, Indonesia
e-mail: suryanto_blppn@yahoo.com

© Springer Nature Singapore Pte Ltd. 2020
S. K. Ghosh (ed.), *Waste Management as Economic Industry Towards Circular Economy*,
https://10.1007/978-981-15-1620-7_5_2

# 1   Waste Management of Balikpapan City

Balikpapan city has a Manggar TPAS with an area of 49.89 ha with the amount of waste in 2017 of 128,932.6 tons of a total population of 636,012 people. The total generation of waste generated is 457.93 tons/day with the composition of organic waste of 58.92% and inorganic waste 41.08%. Of the waste generation, 77.14% was managed in TPAS Manggar amounting to 353, 24 tons/day, 22.20% processed in the infrastructure of waste management facilities at 101.68 tons/day and the remaining 0.66% of untreated waste amounted to 3.01 ton/day.

The Manggar waste final processing site (TPAS) is located at Jalan Proklamasi, Rt 36, Manggar village, east Balikpapan sub-district. At the beginning of its construction, in 1997, the area of TPAS Manggar was 25.1 ha. The construction of TPAS Manggar is part of the Kalimantan urban development project (KUDP) program which began from 1998 to 2002. TPAS Manggar, since January 13, 2002, is operated with a sanitary landfill system in an area of 2.6 ha. To date, the total area of Manggar TPAS is 49.89 ha, which is divided into four zones, namely zone I (2.6 ha), zone II (3 ha), zone IIIA (1.5 ha) and IIIB (0.6 ha) and zone IV (10 ha). Currently, zone I to zone III has been fully utilized. In 2017, the preparation of zone IV has been carried out, according to which the contract will be completed by the end of 2018 (Figs. 1 and 2).

The operational management of TPAS Manggar is under the control of the TPAS Manggar service technical implementation unit (UPTD), whose formation is based on mayor regulation no. 41 of 2012 concerning formation of organizations. Work procedures task and duties of TPAS Manager the technical implementing unit of

**Fig. 1**  TPAS manggar zone

**Fig. 2** TPAS manggar facility

the final processing station at the department of sanitation, city parks and ceme-teries Balikpapan as amended by mayor regulation number 10 of 2016 concerning amendments to mayor's regulation number 41 of 2012 concerning formation of orga-nization, work procedures and job descriptions of technical implementation units of final processing sites at Balikpapan city sanitation, parks and cemetery services. At present, due to changes in the Balikpapan city regional work unit (SKPD), organi-zational structure adjusted to the issuance of government regulation number 18 of 2016 concerning government institutions, and as of January 2017, the TPAS Manggar UPTD is effectively under the Balikpapan city environmental service.

In accordance with the provisions of the legislation, basically the main func-tion of TPAS is merely as a place of disposal and final processing of waste. In its development, the function of TPAS Manggar has been complemented by other activ-ities, including: recycling of organic and inorganic waste, utilization of leachate and methane gas, development of tourism and development of green open space (RTH). The functions of developing the TPAS Manggar will continue to be carried out and developed and utilized more widely for all parties.

The total budget for waste management in Balikpapan city in 2018 is ± rp. 56,000,000,000—includes transportation, sweeping activities, operations of TPAS officers and infrastructure for waste management, and those that enter the regional treasury in the form of waste management fees are ± rp. 10,000,000,000. In addition to TPAS Manggar, other facilities and infrastructure owned by the Balikpa-pan city government to support waste management with a circular economy system include 117 waste banks, one unit of ITF with 22 workers, MRF one unit with 28 workers, TPS is 521 units, 3r TPS consists of eight units, compost houses, waste collection and transfer of waste depots are three units, garbage transport fleet is 72 units and support for adequate human resources is 712 people.

From all waste processing facilities and infrastructure, the results of the waste management produced in the form of compost from organic waste in the ITF amounted to 6.5 tons/day, from 3r TPS of 0.97 tons/day and from the compost house of 25.81 tons/day, whereas inorganic waste managed at MRF is 3.89 tons/day and from waste banks is 2.77 tons/day.

Whereas in the framework of handling coastal and marine waste generation, the city government has obtained one compactor unit from the special allocation fund (DAK). At this time, the procurement of a 45 m floating cube that will be installed in the great Klandasan River (Ampal River) has also been carried out. And in the next planning, it has proposed activities through DAK and financial assistance, adding floating cube 900 m long.

In addition to the aforementioned efforts, other efforts carried out by the city government in order to minimize waste generation on the coast and sea are by adding other supporting facilities and infrastructure in the 2018 fiscal year, among others:

1. Procurement of 13 units of three-wheeled vehicles;
2. Procurement of pickup vehicles;
3. Activating waste banks and waste sorting houses in coastal areas.

In addition to the infrastructure and facilities mentioned above, for waste management in the area, the Balikpapan city government also recruited and financed as many as 30 fishermen to clean up trash in the area in a boat and six wheelbarrows, with the task area covering 20 fishermen serving in the Damai village area to Kampung Baru and 10 fishermen in the Damai village area to Prapatan village.

## 2   Government Cooperation

There needs to be a strategy in the development of the city ecology concept, namely zero waste to landfill, through reducing and managing waste from its source. Through these activities, it is expected that the waste transported to landfill can be significantly reduced and lead to an increase in the age of landfill utilization.

In addition to this, the city government has carried out cooperation with various agencies such as JICA, PT. Fertilizer Indonesia, and the plan of government and business cooperation (PPP) for waste management in TPAS Manggar.

The Balikpapan city government's collaboration with JICA began in 2014 until 2017, and in this period, many efforts have been made which are as follows:

1. Socialization of waste sorting from sources to the Gunung Bahagia village community
2. Inventory of Balikpapan city waste composition
3. Pilot the project of sorting waste from sources in Gunung Bahagia village

Waste sorting activities have begun in 2014 in Gunung Bahagia village, south Balikpapan district. In the initial phase, this activity involved 13 RTS with a population of ±3245 people and 696 households.

At present, the sorting activities are carried out on 57 RTS with a total population of 17,956 people with 5248 households and will subsequently be applied to the entire community of Balikpapan city with the legal basis of Balikpapan city regulation number 13 of 2015 concerning household waste management and household-like waste.

In the waste sorting and processing program from the source, residents do not immediately dispose of garbage into the TPS, but are obliged to first sort waste from sources/homes. The waste that has been disaggregated is then taken to the trash stop, which is a temporary place to dispose of garbage at the designated hour, which is 7:00 a.m. to 9:00 a.m. The schedule for transporting garbage for organic waste is carried out five days a week, for example, on Monday, Tuesday, Wednesday, Friday and Saturday. However, inorganic waste is only disposed of one day every week, that is, on Thursday. As such, no garbage disposal is carried out on Sundays.

Disaggregated rubbish has been collected by residents at the garbage stop, and then processing will be carried out, namely organic waste from the trash stop will be transported to the transfer depot then to the intermediate stream facility (ITF) and processed into compost, while inorganic waste from the trash stop will be transported by officers to the material recovery facility (MRF), which will then be carried out in an economic transaction with third parties. In addition to MRF, inorganic wastes will also be managed directly by the community through waste banks which are managed directly/independently by community groups (Fig. 3).

**Fig. 3** Balikpapan new waste management flow

The cooperation between Balikpapan city government and PT. Indonesian fertilizers have been started since 2016, with the main focus being the production of organic granule fertilizer from raw materials for organic waste and chip mill industry waste. PT. Pupuk Indonesia has carried out a series of researches, trials, laboratory tests and feasibility studies on feasibility plans for the production of granular organic fertilizers in the city of Balikpapan. The raw material for organic fertilizer is obtained from organic waste in the city of Balikpapan and mixed with the chip mill, which is got from industrial waste from PT. Kutai chip Mill. From this trial plan, I hope there will be great benefits for the people of Balikpapan city, namely the management of organic waste and the availability of organic fertilizers that can be used as an alternative to chemical fertilizers.

Until now, it is being considered the most appropriate location to establish the granule organic fertilizer production unit. Land requirements are $\pm$ 20,000 m$^2$ (2 ha). Some proposed locations such as in TPAS Manggar, at ITF and in other locations.

## 3   City Policies and Regulations

In the framework of sustainable waste management, the Balikpapan city government has established several policies, including the stipulation of regional regulation number 13 of 2015 concerning household waste management and household-like waste, as well as mayor regulation number 8 of 2018 concerning reduction in the use of plastic bags. This is an effort to support the achievement of waste processing targets from sources of 25.80% up to 2021 from all waste generated by community activities in Balikpapan city.

## 4   Planning Program

The Balikpapan city government has an effort to manage organic waste originating from Pandan sari market activities plans to utilize the organic waste into compost that will be carried out in Kampung Atas Air Margasari and Pandan sari market.

Waste management in Pandan sari market is not only used as compost but also used for methane gas from organic waste that can be used as fuel and electrical energy street lighting around the location.

With various policies that have been established as waste management efforts, it is expected that the implementation of waste management from sources can be successful. Furthermore, the city government plans to develop a city-scale recycling center by first conducting a feasibility study.

# The Beneficiation of Waste as Part of the Implementation of the Circular Economy in South Africa

A. S. Pillay and C. A. Pillay

**Abstract** Waste management is a global challenge. The circular economy is now gaining momentum worldwide. It is a concept that even makes complete sense to ordinary people. It shifts from a traditional linear model to an innovative circular model. This embraces the new thinking of the cradle-to-cradle approach.

**Keywords** International Society of Waste Management, Air and Water

## 1 Introduction

Waste management is a global challenge. The circular economy is now gaining momentum worldwide. It is a concept that even makes complete sense to ordinary people. It shifts from a traditional linear model to an innovative circular model. This embraces the new thinking of the cradle-to-cradle approach (Abubakar 2018).

Pasqua Heard in the article on "Three circular economy projects in South Africa" discusses the concept of circular economy as follows "While discussion about the circular economy began in the 1970s, it is a relatively new environmental term for South Africa. In a nutshell, the circular economy, otherwise known as the closed-loop economy, aims to reduce pollution through a process whereby resources and products are reused before they reach the waste stage. It counters a linear economy approach which is based on making, using and disposing products. The circular economy aims to use as little energy as possible, so, ideally, products would be used and reused in their natural state" (Heard 2016).

According to the South African baseline study conducted in 2011, approximately 10% of waste was recycled and 90% was landfilled (Department of Environmental Affairs 2012).

A. S. Pillay
Waste Policy and Information Management, Pretoria, South Africa
e-mail: apillay@environment.gov.za

C. A. Pillay (✉)
University of Pretoria, Pretoria, South Africa
e-mail: chislonpillay@icloud.com

© Springer Nature Singapore Pte Ltd. 2020
S. K. Ghosh (ed.), *Waste Management as Economic Industry Towards Circular Economy*,
https://10.1007/978-981-15-1620-7_5_3

South Africa produces electricity primarily from fossil fuel, coal burning. This results in large generation of ash from these processes (approximately 37 million tons/annum). South Africa is also a large producer of iron and steel products. This process thus generates a large quantity of slag (approximately 5 million tons/annum). South Africa also generates a large quantity of gypsum and biomass (approximately 15 million tons/annum).

The National Environmental Management Waste Act, Act 59 of 2008 (NEMWA) permits the minister to make regulations to ensure proper administration of the Act. The minister has considered making regulations to exclude a waste stream or portion of a waste stream from the definition of waste.

## 2 Currently

According to the National Environmental Management Waste Act, Act 59 of 2008 (NEMWA) (South African Government 2014) Waste is defined as:

(a) any substance, material or object, that is unwanted, rejected, abandoned, discarded or disposed of, or that is intended or required to be discarded or disposed of, by the holder of that substance, material or object, whether or not such substance, material or object can be reused, recycled or recovered and includes all wastes as defined in Schedule 3 to this act; or

(b) any other substance, material or object that is not included in Schedule 3 that may be defined as a waste by the minister by notice in the Gazette, but any waste or portion of waste, referred to in paragraphs (a) and (b), ceases to be a waste

   (i)   once an application for its reuse, recycling or recovery has been approved or, after such approval, once it is, or has been reused, recycled or recovered;
   (ii)  where approval is not required, once a waste is, or has been reused, recycled or recovered;
   (iii) where the minister has, in terms of Section 74, exempted any waste or a portion of waste generated by a particular process from the definition of waste; or
   (iv)  where the minister has, in the prescribed manner, excluded any waste stream or a portion of a waste stream from the definition of waste.

The exclusion of waste or portion of waste from the definition of waste will enable the recycling of many waste materials for beneficial purposes.

The regulations have been developed to provide the criteria and requirements to exclude a waste stream or a portion of a waste stream from the definition of waste for beneficial purposes.

This will positively contribute to job creation, stimulation of the green economy, and diversion of waste away from landfilling towards reuse, recycle and recovery as contained in the waste hierarchy.

## 3   Literature Review/Survey

OECD (2007) used the working definition for environmentally sound management as follows:

> a scheme for ensuring that wastes and used and scrap materials are managed in a manner that will save natural resources, and protect human health and the environment against adverse effects that may result from such wastes and materials.

The OECD Council adopted this working definition and included it in its recommendations on environmentally sound management of waste. There were three specific objectives on waste.

1.  "sustainable use of natural resources, minimisation of waste and protection of human health and the environment from adverse effects that may result from waste;
2.  fair competition between enterprises throughout the OECD area through the implementation of 'core performance elements' (CPEs) by waste management facilities, thus contributing to a level playing field of high environmental standards;
3.  through incentives and measures, diversion of waste streams to the extent possible from facilities operating with low standards to facilities that manage waste in an environmentally sound and economically efficient manner".

## 4   Approach

### 4.1   Internationally

Beneficiation is a common practice internationally. Fly ash and boiler ash are currently being used as an integral part of the cement manufacturing process (Pacheco-Torgal 2017). This process demonstrated that there were little or no significant negative impacts from the use of these products in the manufacture of cement.

Gypsum has been used for the manufacture of portioning boards and ceiling boards for years. Waste tyres have been used as an alternative fuel source in cement kilns (Hartley et al. 2016). Various other materials have been used in other applications.

### 4.2   South African Approach

South Africa has published the Waste Exclusion Regulations, 2018 were published on 18 July 2018 for implementation. South Africa is in the process of implementing the circular economy approach in waste.

South Africa seeks to embrace the opportunities contained in this approach to achieve the goals of the National Development Plan and also to address the triple challenge of poverty alleviation, creation of jobs and stimulating the economy which government is facing (McLellan et al. 2018).

The existing legislative regime stipulates that certain waste management activities require a waste management licence in terms of Section 19 of the National Environmental Management Waste Act, Act 59 of 2008 (NEMWA).

The following waste management activities require a licence:

- Storage and transfer of waste;
- Recycling and recovery;
- Treatment of waste;
- Disposal of waste on land; and
- Storage, treatment and processing of animal waste.

The four waste streams namely ash, slag, biomass and gypsum will be recycled and hence will trigger an activity requiring a waste management licence. These regulations provide relief in that the requirement for a waste management licence for recycling these four waste streams would be excluded.

This removes the burden of following an Environmental Impact Assessment process (EIA) and then lodging an application for a waste management licence. These processes normally require the services of a registered Environmental Assessment Practitioner (EAP). There are exorbitant costs attached to the development of an EIA and lodging of a waste management licence application. These regulations effectively remove these costs.

These prescribed administrative processes are bound by regulated timeframes and often result in "missed opportunities" due to the long timelines. The time for an EIA is 300 days provided there is no additional information required. Any additional information requests may result in significant delays in processing the applications within the prescribed timeframes. This demonstrates the "missed opportunities".

The costs of these processes are a barrier to entry to many entrepreneurs resulting in many innovative and creative ideas not being implemented. Many funding institutions refuse projects by emerging small businesses and entrepreneurs due to the uncertainty of the approval processes.

South Africa and its government are seeking opportunities to remove unnecessary hurdles to promote, the diversion of waste away from landfill towards recusing, recycling and recovery, the development of small businesses and developing entrepreneurs (Dasgupta 2014).

The regulations prescribe the criteria for considering a waste stream for exclusion from the definition of waste. The regulations prescribe the following:

(a) the applicant demonstrates that the waste is being or has been or will be used for a beneficial purpose either locally or internationally;
(b) the applicant submits a risk assessment demonstrating that the intended beneficial use of the excluded waste can be managed in such a way as to ensure that

the intended beneficial use will not result in significant adverse impacts on the environment; and

(c)  a risk management plan responding to the risks identified in the risk assessment undertaken in terms of paragraph (b) above accompanies any delivery of the excluded waste to the user.

# 5  Demonstrating Beneficial Use

The regulations require some documented evidence of the waste material being used for beneficial purposes without significant negative impacts to the environment or health and well-being. With the international arena already using many waste materials for alternative purposes, it simply needs to be referenced. The examples cited above such as boiler ash, gypsum and slag are currently being widely used for numerous applications successfully (Organisation for Economic Co-operation and Development 2013).

# 6  Risk Assessment

The risk assessment must contain the following information:

(a)  provide information that is facility based;
(b)  description and source of the waste;
(c)  intended uses of the excluded waste;
(d)  description of the methodology used to assess the hazardous characteristics of the waste that is to be excluded;
(e)  identification of any potential risks relating to all the activities associated with the intended beneficial use of the excluded waste; and
(f)  identification of mitigation measures that can be used to manage the risks identified in paragraph (e) above.

# 7  Risk Management Plan

The risk management plan must include the following:

(a)  where the material is classified as hazardous, a Safety Data Sheet which complies with the requirements of SANS 10234;
(b)  permitted uses for which the waste material may be used; and

(c) a mechanism to record the amount of waste distributed to specific users for a permitted use; including the number of enterprises established or supported and the extent to which previously disadvantaged individuals have been supported.

# 8  Methodology

The regulations were developed in consultation with industry and other relevant government departments such as Trade and Industry, Economic Development and Small Businesses.

The Draft Regulations were published for public comment for 30 days in two national newspapers. Several comments and inputs were received from 21 institutions. These comments sought to refine the regulations. The comments and inputs were incorporated into the regulations and then finalised and published for implementation.

# 9  Waste Generation Data

The department conducted a baseline study during 2011 and Tables 1, 2, 3 and 4 are the results of the study as adapted.

# 10  Implementation of the Waste Hierarchy

These regulations seek to give effect to the waste hierarchy in its entirety. The idea is to firstly consider reduction at source, reusing, recycling, recovery and lastly disposal (Department of Environmental Affairs, 2017).

# 11  Reduction

There have been many initiatives with regard to reduction at source. There was a huge drive to optimise the process and improve efficiencies with manufacturing processes. Many facilities demonstrated successes in reducing their waste generation.

**Table 1** 2011 general waste by management option

| General waste 2011 | | Generated | Recycled | Landfilled | Recycled |
|---|---|---|---|---|---|
| | | Tonnes | | | % |
| GW01 | Municipal waste (non-recyclable portion) | 8,062,934 | – | 8,062,934 | 0 |
| GW10 | Commercial and industrial waste | 4,233,040 | 3,259,441 | 973,599 | 77 |
| GW13 | Brine | See Table 4 | | | |
| GW14 | Fly ash and dust from miscellaneous filter sources | See Table 4 | | | |
| GW15 | Bottom ash | See Table 4 | | | |
| GW16 | Slag | See Table 4 | | | |
| GW17 | Mineral waste | See Table 4 | | | |
| GW18 | Waste of electric and electronic equipment (WEEE) | See Table 4 | | | |
| GW20 | Organic waste | 3,023,600 | 1,058,260 | 1,965,340 | 35 |
| GW21 | Sewage sludge | See Table 4 | | | |
| GW30 | Construction and demolition waste | 4,725,542 | 756,087 | 3,969,455 | 16 |
| GW50 | Paper | 1,734,411 | 988,614 | 745,797 | 57 |
| GW51 | Plastic | 1,308,637 | 235,555 | 1,073,082 | 18 |
| GW52 | Glass | 959,816 | 307,141 | 652,675 | 32 |
| GW53 | Metals | 3,121,203 | 2,496,962 | 624,241 | 80 |
| GW54 | Tyres | 246,631 | 9865 | 236,766 | 4 |
| GW99 | Other | 36,171,127 | – | 36,171 127 | 0 |
| Total general waste [T] | | 59,353,901 | 5,852,484 | 53,501,417 | ~10 |

## 12 Reusing

There have been some efforts in this area of reusing. South Africa has developed an Industrial Symbiosis programme in three provinces and has made some significant achievements in reusing waste materials. The programme is supported by government and also unlocks many barriers to enable such initiatives. The programme identifies waste streams that can be used as a resource for other industrial applications and facilitates the processes of linking the relevant facilities together to form a mutually beneficial relationship. The Exclusion Regulations seek to improve our performance in the reusing step of the waste hierarchy.

**Table 2** 2011 hazardous waste by management option

| Hazardous waste | | Generated | Recycled | Treated | Landfilled | Recycled |
|---|---|---|---|---|---|---|
| | | Tonnes | | | | % |
| HW01 | Gaseous waste | 55 | – | 55 | – | – |
| HW02 | Mercury-containing waste | 868 | – | – | 868 | – |
| HW03 | Batteries | 32,912 | 32,254 | – | 658 | 98 |
| HW04 | POP waste | 486 | – | 80 | 406 | – |
| HW05 | Inorganic waste | 290,154 | – | – | 290,154 | – |
| HW06 | Asbestos-containing waste | 33,269 | – | – | 33,269 | – |
| HW07 | Waste oils | 120,000 | 52,800 | – | 67,200 | 44 |
| HW08 | Organic halogenated and/or sulphur-containing solvents | 111 | – | – | 111 | – |
| HW09 | Organic halogenated and/or sulphur-containing waste | 8389 | – | 64 | 8325 | – |
| HW10 | Organic solvents without halogens and sulphur | 771 | – | – | 771 | – |
| HW11 | Other organic waste without halogen or sulphur | 202,708 | – | – | 202,708 | – |
| HW12 | Tarry and bituminous waste | 255,832 | – | – | 255,832 | – |
| HW19 | Healthcare risk waste | 46,291 | – | 46 291 | – | – |
| HW99 | Miscellaneous | 327,250 | – | – | 327,250 | – |
| Total hazardous [T] | | 1,319,096 | 85,054 | 46,490 | 1,187,552 | ~6 |

# 13   Recycling

We as a country have embarked on several recycling initiatives. Plastics, paper, metals, tyres and many more waste streams have been targeted for recycling. Year on year we see major improvement in the recycling rates of these waste streams.

**Table 3** SAWIS data reported per waste stream

| Code | Waste reference | Disposal | Recovery/recycling | Waste treatment | Total |
|---|---|---|---|---|---|
| GW01 | Municipal waste | 7,198,143 | 11,029 | 55,518 | 726,4690 |
| GW10 | Commercial and industrial waste | 501,648 | 4848 | 4005 | 510,501 |
| GW13 & HW13 | Brine | 2086 | 0 | | 2086 |
| GW14 & HW14 | Fly ash and dust from miscellaneous filter sources | 154,176 | 0 | 0 | 154,176 |
| GW15 & HW15 | Bottom ash | 5,561,370 | 209 | 0 | 5,561,370 |
| GW16 & HW16 | Slag | 1,003,054 | 52,107 | 0 | 1,055,161 |
| GW17 & HW17 | Mineral waste | 7370 | 3334 | 0 | 10,704 |
| GW18 & HW18 | Waste of electric and electronic equipment (WEEE) | 47 | 133,355 | 10,743 | 144,145 |
| GW20 | Organic waste | 999,052 | 371,088 | 159,202 | 1,529,342 |
| GW21 | Sewage sludge | 4475 | 0 | | 4475 |
| GW30 | Construction and demolition waste | 940,762 | 132,990 | | 1,073,752 |
| GW50 | Paper | 254 | 1,392,060 | 0 | 1,392,528 |
| GW51 | Plastic | 11,310 | 195,886 | 0 | 207,196 |
| GW52 | Glass | 191 | 72,505 | | 72,696 |
| GW53 | Metals | 43 | 536,359 | 0 | 536,402 |
| GW54 | Tire's | 564 | 3385 | 0 | 3949 |
| GW99 | Other | 261,408 | 26,445 | 44 | 287,897 |
| HW01 | Gaseous waste | 0 | | 200 | 200 |
| HW02 | Mercury-containing waste | 530 | 0 | 0 | 530 |
| HW03 | Batteries | 10 | 0 | 23,882 | 23,892 |
| HW04 | POP Waste | 228 | 0 | 0 | 228 |
| HW05 | Inorganic waste | 242,006 | 0 | 13,129 | 255,135 |
| HW06 | Asbestos-containing waste | 7838 | | | 7838 |

(continued)

**Table 3** (continued)

| Code | Waste reference | Disposal | Recovery/recycling | Waste treatment | Total |
|---|---|---|---|---|---|
| HW07 | Waste Oils | 3713 | 12,032 | 106,776 | 122,521 |
| HW08 | Organic halogenated and/or sulphur-containing solvents | 1308 | 140 | 2952 | 4400 |
| HW09 | Organic halogenated and/or sulphur-containing waste | 2721 | 0 | 0 | 2721 |
| HW11 | Organic solvents without halogens and sulphur | 169,695 | 1,242,314 | 46,759 | 1,458,768 |
| HW12 | Other organic waste without halogen or sulphur | 5215 | 0 | 0 | 5215 |
| HW19 | Tarry and bituminous waste | 2998 | 0 | 297,784 | 300,782 |
| HW20 | Healthcare risk waste | 2380 | | | 2380 |
| HW99 | Miscellaneous | 83,604 | 104 | 580 | 84,541 |
| Total | | 17,172,493 | 4,190,190 | 721,574 | 22,084,724 |

**Table 4** Percentage of reporting waste management activities per province

| | Disposal (%) | Recovery/recycling (%) | Treatment (%) | Total (%) |
|---|---|---|---|---|
| Eastern Cape | 37.74 | 12.96 | 45.45 | 27.12 |
| Free State | 30.16 | 18.75 | 33.33 | 29.27 |
| Gauteng | 68.89 | 37.89 | 36.36 | 43.20 |
| Kwazulu-Natal | 35.48 | 20.29 | 38.10 | 27.15 |
| Limpopo | 54.35 | 12.82 | 70.00 | 38.95 |
| Mpumalanga | 45.33 | 18.37 | 50.00 | 38.36 |
| North West | 38.00 | 25.00 | 42.86 | 34.12 |
| Northern Cape | 70.45 | 21.43 | 66.67 | 59.02 |
| Western Cape | 45.81 | 53.33 | 32.14 | 47.57 |
| Total | 45.87 | 29.64 | 41.61 | 38.56 |

# 14   Recovery

South Africa has a few incineration plants or waste to energy facilities. There are some successes in this area however more needs to be accomplished. There needs to be a market developed for the off-takes for the products from these processes. There is significant room for research and development in this space.

# 15 Disposal

We have excelled in this area of the hierarchy when one considers our 2011 Baseline study wherein we established that 90% of our waste is disposed of at landfills. We are exploring innovative ways to make a paradigm shift in this area.

The following list provides the list of waste streams currently applied for, beneficial use, impacts and mitigation measures for the negative impacts.

(1) **Description of Waste: Waste Slag from Ferrochrome Metallurgy**

| Prescribed use of waste | Intermediate processes required to make waste available for use | Potential negative or positive environmental impacts of use | Mitigation of negative impacts |
| --- | --- | --- | --- |
| (a) Use as aggregates | Recovery from disposal site | Negative: Potential dust emission associated with handling of aggregate material | Manage and mitigate dust emissions within occupational health standards and comply with the requirements of the National Environmental Management: Air Quality Act, 2004 (Act No. 39 of 2004) and any other relevant legislation |
| (b) Concrete aggregates | Crushing and screening | Positive: Eliminate the need for virgin aggregate material to be mined with its associated environmental impact | None |
| (c) Road base and covering and road stabilisation | None | Positive: Eliminate the need for virgin aggregate material to be mined with its associated environmental impact | None |

(continued)

(continued)

| Prescribed use of waste | Intermediate processes required to make waste available for use | Potential negative or positive environmental impacts of use | Mitigation of negative impacts |
|---|---|---|---|
| (d) Asphaltic concrete and other bituminous mixtures | None | Positive: Eliminate the need for virgin aggregate material to be mined with its associated environmental impact | None |
| (e) Construction fill | None | Positive: Eliminate the need for virgin aggregate material to be mined with its associated environmental impact | None |
| (f) Concrete products | None | Positive: Eliminate the need for virgin aggregate material to be mined with its associated environmental impact | None |
| (g) Plaster and gunite sands | None | Positive: Eliminate the need for virgin aggregate material to be mined with its associated environmental impact | None |
| (h) Railroad ballast | None | Positive: Eliminate the need for virgin aggregate material to be mined with its associated environmental impact | None |
| (i) Roofing granules | None | Positive: Eliminate the need for virgin aggregate material to be mined with its associated environmental impact | None |

(continued)

(continued)

| Prescribed use of waste | Intermediate processes required to make waste available for use | Potential negative or positive environmental impacts of use | Mitigation of negative impacts |
|---|---|---|---|
| (j) Filtration media | None | Positive: Eliminate the need for virgin aggregate material to be mined with its associated environmental impact | None |
| (k) Pipe filling material | None | Positive: Eliminate the need for virgin aggregate material to be mined with its associated environmental impact | None |
| (l) Backfilling | None | Positive: Eliminate the need for virgin aggregate material to be mined with its associated environmental impact | None |
| (m) Dam construction and stabilisation material | None | Positive: Eliminate the need for virgin aggregate material to be mined with its associated environmental impact | None |
| (n) Construction of drainage systems | None | Positive: Eliminate the need for virgin aggregate material to be mined with its associated environmental impact | None |
| (o) Hydroponic filling material | None | Positive: Eliminate the need for virgin aggregate material to be mined with its associated environmental impact | None |

(continued)

(continued)

| Prescribed use of waste | Intermediate processes required to make waste available for use | Potential negative or positive environmental impacts of use | Mitigation of negative impacts |
|---|---|---|---|
| (p) Production of cement | None | Positive: Eliminate the need for virgin material to be mined with its associated environmental impact | None |

## (2) Description of Waste: Ash from Combustion Plants

| Prescribed use of waste | Intermediate processes required to make waste available for use | Potential negative or positive environmental impacts of use | Mitigation of negative impacts |
|---|---|---|---|
| (a) Brick making | None | **Positive**: Reduced landfill waste. Ash bound in cement/clay therefore dust and potential leachate impacts are minimised | None |
|  |  | **Negative**: Handling | Duty of care principles provided for in the National Environmental Management Act, 1998 (Act No. 107 of 1998) to be followed by supplier which includes ensuring correct storage facility on user site (Norms & Standards for Storage of Waste, 2013 to be adhered to); dust control; storm water management |

(continued)

(continued)

| Prescribed use of waste | Intermediate processes required to make waste available for use | Potential negative or positive environmental impacts of use | Mitigation of negative impacts |
|---|---|---|---|
| (b) Block making | None | **Positive**: Reduced landfill waste Ash bound in cement therefore dust and potential leachate impacts are minimised | None |
| | | **Negative**: Handling | Duty of care principles provided for in the National Environmental Management Act, 1998 (Act No. 107 of 1998) to be followed by supplier which includes ensuring correct storage facility on user site (Norms & Standards for Storage of Waste, 2013 to be adhered to); dust control; storm water management |
| (c) Production of cement | None | **Positive**: Improved material use/resource efficiency Reduced landfill waste | None |
| | | **Negative**: None | None |
| (d) Landfill capping | Mixing | **Positive**: Ash mixed with fibre sludge has proven effective as alternative to soil being used for capping purposes | Mixing to be done on a licensed landfill site |
| | | **Negative**: Dust generation | |
| (e) Backfill in old mine workings | None | **Negative**: Spillages | None |

(continued)

(continued)

| Prescribed use of waste | Intermediate processes required to make waste available for use | Potential negative or positive environmental impacts of use | Mitigation of negative impacts |
|---|---|---|---|
| | | **Positive**: Beneficial use of material | None |
| (f) Inorganic fertilizer | None | **Positive**: Return nutrients and micro-elements to soil (minimise the need for chemical fertilizers to be applied). Soil conditioning | None |
| (g) Soil ameliorant | None | **Positive**: Return nutrients and micro-elements to soil (minimise the need for chemical fertilizers to be applied). Soil conditioning | None |
| (h) Asphalt and other bituminous mixtures | None | **Positive**: Improved material use/resource efficiency. Reduced waste to landfill | None |
| (i) Road construction | None | Negative: Dust generation | Manage and mitigate dust emissions within occupational health standards and comply with the requirements of the National Environmental Management: Air Quality Act, 2004 (Act No. 39 of 2004) and any other relevant legislation |
| (j) Foundations | None | None | None |

(continued)

(continued)

| Prescribed use of waste | Intermediate processes required to make waste available for use | Potential negative or positive environmental impacts of use | Mitigation of negative impacts |
|---|---|---|---|
| (k) Bulking agent for compositing | None | Negative: Dust generation | Manage and mitigate dust emissions within occupational health standards and comply with the requirements of the National Environmental Management: Air Quality Act, 2004 (Act No. 39 of 2004) and any other relevant legislation |

(3) **Description of Waste: Gypsum from Pulp, Paper and Cardboard Production and Processing**

| Prescribed use of waste | Intermediate processes required to make waste available for use | Potential negative or positive environmental impacts of use | Mitigation of negative impacts |
|---|---|---|---|
| (a) Soil conditioner | None | **Positive:** Reduced landfill waste. Positive impact on sodium absorption ratio in soil | None |
| (b) Inert products (such as board manufacturing) | None | **Positive:** Improved material use/resource efficiency Reduced landfill waste | None |

(4)  **Description of Waste: Biomass (Bark, Offcuts, Sawdust) from Pulp, Paper and Cardboard Production and Processing**

| Prescribed use of waste | Intermediate processes required to make waste available for use | Potential negative or positive environmental impacts of use | Mitigation of negative impacts |
|---|---|---|---|
| (a) Composting | Composting in windrows | **Positive**: Enhanced soil fertility. Reduced landfill waste Job creation | None |
| | | **Negative**: Possible leachate | None |
| (b) Soil conditioner in plantations | Direct to field | **Positive**: Improved soil properties Reduced landfill waste Job creation | None |
| | | **Negative**: Possible concentration of contaminants | Assessment of correct application volumes based on soil properties prior to application |
| (c) Animal bedding | Bagged | **Positive**: Reduced landfill waste | None |

# 16  Job Creation, Development of Small Businesses and Capacity Building

This is a priority for the Government of South Africa. Along with a high unemployment rate of approximately 27.2% come various social challenges. One of these is a hike in the crime rate which ultimately affects the economic growth within South Africa.

Hence the government is looking at various opportunities to provide meaningful employment to the youth of the country. It seeks to fast-track the development of small business and create a number of jobs. Government is removing as many barriers to entry as possible to promote activities and businesses that create jobs. This has been the commitment of government and many processes and procedures are being streamlined to support this initiative.

# 17  Conclusions

This intervention seeks to address the following challenges: Rapid increase in waste generation, Rapid depletion of landfill airspace, Stagnant recycling rates and increasing the diversion of waste away from landfill towards recycling, reusing and recovery.

Industry is ecstatic for this initiative and is working closely with the government to ensure that this opportunity is capitalised upon by all stakeholders. These regulations have been published for implementation and the applications for the four streams have been submitted and are in the process of being evaluated with a recommendation to the minister for approval.

This Regulatory initiative is estimated to divert approximately 15 million tons of waste per annum away from landfill towards reuse, recycling and recovery. It has the potential to create approximately 10,000 direct jobs over a five-year implementation period.

# References

Abubakar, F. (2018). *White Rose eThesis Online*. [Online] Available at: http://etheses.whiterose.ac.uk/20947/. Accessed August 2, 2018.

Dasgupta, T. (2014). Composition of municipal solid waste generation and recycling scenario of building materials. *International Journal of Scientific Engineering and Technology, 3*(10), 1259–1264.

Department of Environmental Affairs. (2012). *National waste information baseline report*. Pretoria: Department of Environmental Affairs.

Department of Environmental Affairs. (2017). *Industry waste plans presentation*. [Online] Available at: http://pmg-assets.s3-website-eu-west-1.amazonaws.com/170314Industry_Waste.pdf. Accessed August 2, 2018.

Hartley, F., Caetano, T. & Daniels, R. C. (2016). *University of Cape Town*. [Online] Available at: http://webcms.uct.ac.za/sites/default/files/image_tool/images/119/Papers-2016/16-Hartley-etal-waste_tyres.pdf. Accessed August 2, 2018.

Heard, P. (2016). *Bizcommunity*. [Online] Available at: http://www.bizcommunity.com/Article/196/750/149086.html. Accessed August 17, 2018.

McLellan, B., Corder, G., & Ali, S. (2018). *Sustainability of Rare Earths—An Overview of the State of Knowledge*.

Organisation for Economic Co-operation and Development. (2013). OECD. [Online] Available at: http://www.oecd.org/env/waste/OECD%20Work%20on%20SMM_update%2011-02-2013.pdf. Accessed August 2, 2018.

Pacheco-Torgal, F. (2017). *Handbook of recycled concrete and demolition waste*. Cambridge: Woodhead Publishing.

South African Government. (2014). Act No. 26 of 2014: National Environmental Management: Waste Amendment Act. Government Gazette.

# Determinants of Sustainable New Product Development and Their Impacts in Manufacturing Companies

**Sudeshna Roy, Nipu Modak and Pranab K. Dan**

**Abstract** Necessity of sustainable new product development (S-NPD) has becoming increasingly relevant in the present era for sustaining in the global competition. Though S-NPD has been neglected earlier, its vital role in business and academic perspectives indulges the companies to be involved in S-NPD for their own sake. This research identifies seven critical success factors (CSFs) of S-NPD and their indicators as well. After recognition, it realizes the importance for implementation of these factors in Indian manufacturing companies. It recognizes the CSFs such as structural configuration, learning practice, strategic configuration, internal perspectives, external issues, PLC analysis and additional performance for S-NPD. This S-NPD can be achieved by effort to reduce cost and increase profitability, achieving resource efficiency, customer satisfaction and reduction of environmental pollution created by the product, health and safety aspects, social aspects and life cycle analysis. This study accumulates the primary data from 255 manufacturing experts mainly from design and development team for data analysis. The structural equation modeling (SEM) approach is been employed to analyze the combined impact of these CSFs on sustainable NPD by using IBM SPSS AMOS 21.0 software. This study interprets that all the CSFs have positive impact on sustainable product development for enhancing the S-NPD. Strategic configuration has been identified as the most impacted success factor for S-NPD. Among success measures, customer satisfaction is recognized as the most vital measure followed by health and safety aspects, life cycle analysis, social aspects, resource efficiency, reduction of environmental pollution created by the product and reduced cost and increased profitability. This empirical research helps to draw the managerial implications for highlighting the success factors and measures as per their importance for S-NPD.

**Keywords** Sustainable new product development (S-NPD) · Critical success factors (CSFs) · Structural equation modeling (SEM) · International society of waste management · Air and water

S. Roy (✉) · N. Modak
Mechanical Engineering Department, Jadavpur University, Kolkata, India
e-mail: sudeshnaroy689@gmail.com

P. K. Dan
Rajendra Mishra School of Engineering Entrepreneurship, IIT Kharagpur, Kharagpur, India

© Springer Nature Singapore Pte Ltd. 2020
S. K. Ghosh (ed.), *Waste Management as Economic Industry Towards Circular Economy*,
https://10.1007/978-981-15-1620-7_5_4

# 1   Introduction

New product development (NPD) is a process of introducing products completely new to market through series of activities for fulfilling the customers' need (Booz 1982). NPD influences cost, quality, development time, customer satisfaction and financial performance of the firm for achieving industrial sustainability. The idea of sustainability is incorporated with NPD focusing on sustainable new product development (S-NPD) by involving sustainability with each phase of NPD to value their customers (Schaltegger 2011; Paramanathan et al. 2004). Environmental issues add a different dimension to NPD process of the firm which is often been neglected. Cleaner production and eco-innovation have been introduced, but these activities are remained as a 'term' for the small and medium-scaled enterprises (SMEs) (Schaltegger and Wagner 2011). In this era of globalization, the NPD is not only an affair of developing something innovative, but to produce new products by considering its adverse effect on environment (Hansen et al. 2009). Environmental hazards have reached an alarming position where each person needs to be concerned about hazardous effects of their consumables. S-NPD is an approach to deliver new products by considering social, economic and environmental aspects together in a single frame (Paech 2007). Association of NPD activities along with the organizational configuration to produce environment-friendly new products is initiated for achieving industrial sustainability (Rennings 2000). The S-NPD activities of large scale and SMEs are different due to their difference in nature, size and innovation attributes (Hillary 2000). The drivers of S-NPD of SMEs are needed to be recognized to facilitate S-NPD for industrial sustainability. These drivers are famously known as critical success factors (CSFs) (Ernst 2002). There are researches identified the CSFs for S-NPD to achieve eco-innovation. Structural configuration, learning practice, strategic configuration, internal perspectives, external issues, product life cycle (PLC) analysis and additional performance are identified as the factors critical to success for S-NPD (de Jesus Pacheco et al. 2017). The combined impact of these variables on S-NPD is essential to measure the performance of the firm in terms of social–economic and environmental aspects. In this scenario, the integrating framework considering the CSFs of S-NPD and the success measures are needed to be developed which is largely unexplored. The consideration of environmental issues for better performance outcome is required to be developed and the essential steps to build the support system and create a conducive ambience for successful implementation of the implications drawn from the analysis.

The objective of the study is to develop an integrative model for realizing the combined impact of environmental aspects along with the organizational and strategic issues of the SMEs for implementing those practices efficiently for achieving desired social, economic and environmental success. The structural model is constructed considering the aforementioned CSFs, and their combined impact on S-NPD is tested by structural equation modeling approach using IBM SPSS AMOS 21.0.

# 2 Research Methodology

## 2.1 Methods

Structural equation modeling (SEM) is a method for representing, estimating and testing the relations between latent variables and their manifests. It is mainly a combination of exploratory factor analysis and multiple regressions (Ullman 2001). Latent variables are those which cannot be measured directly. Manifest variables of the latent construct are the observed or measured variables through which the latent can be measured. SEM comprises two models, namely measurement model and structural model. In measurement model formation, the confirmatory factor analysis (CFA) is performed. The structural model represents the interrelation among the latent constructs and observed variables (Schreiber et al. 2006). This study identifies the role of factors of S-NPD for successful development of sustainable new products in terms of social, economic and environmental aspects. The analysis of the developed framework is performed on the basis of primary data collected from 263 experts of Indian manufacturing industries. The reliability of the accumulated primary data has been tested by using composite reliability (CR) and Cronbach's alpha reliability testing using IBM SPSS AMOS 21.0. The average variance extracted (AVE) is been calculated for testing the discriminant validity of the collected data. The threshold value of CR and AVE is 0.5, whereas in case of $\alpha$ it is 0.8 (Ong et al. 2004). The exploratory factor analysis is performed to measure factor loadings (FLs) to their respective constructs. SEM is employed by using IBM SPSS AMOS 21.0 to develop the structural framework testing the hypotheses developed from the available literature and the experts' opinion. CFA is performed for calculating the standard regression weights (SRWs) of the manifest variables in measurement model section. In case of structural model formation, path estimation between the latent constructs is performed by using maximum likelihood method.

## 2.2 Hypothesis Development

This work involves development of seven hypotheses relating the CSFs to S-NPD for realizing their impact on performance attributes of development of sustainable new products. Structural configuration of SMEs is one of the critical components which are to be concerned for developing new products. Organizational structure, methods adopted for NPD, external collaboration with suppliers and customers, R&D activities, risk management and moreover managerial support are the essential criteria each firm must be concerned for successful NPD (Aykol and Leonidou 2014; Ackermann and Eden 2011). Learning practice is another constituent essential for S-NPD activities. It includes proper training in both internal and external bases, practice to cooperate with extern stakeholders and strong technological background with advisory committee to develop sustainable new products (Ackermann and Eden 2011;

Blackburn 2007). Strategic configuration is another constituent of the firm which develops the strategy of success. Managerial insight is the most vital indicator of the strategy development. It also comprises strong synchronization among internal teams and variations in strategies as per market changes occurred (Boly et al. 2014; Blackburn 2008). Besides these organizational and strategic issues, product life cycle (PLC) analysis is essential to associate environmental aspects for S-NPD. The awareness about the raw materials having hazardous impacts on environment, usage of recycled materials, recycling of scrap metals along with all valuable components and optimized design of the new product to be developed (Ackermann and Eden 2011; Blackburn 2008; Bertoni et al. 2015). Moreover, there are additional performance aspects like brand image, attraction of customers and employers and lean manufacturing to reduce the waste are also considered as one of the vital CSFs of S-NPD to be taken care of (Ackermann and Eden 2011; Bertoni et al. 2015). The social, economic and environmental issues are considered as the performance attributes of the S-NPD which covers reduction of cost and increment of the profit of the firm, resource efficiency, customer satisfaction, minimization of environmental pollution, health and safety aspects, social aspects and life cycle analysis. The hypotheses developed from the above discussions are listed below. The path model developed from these developed hypotheses is represented in Fig. 1:

H1: Strategic configuration of the firm motivates the S-NPD.
H2: Learning practice enhances S-NPD activities of the firm.
H3: Proper strategic configuration can boost up S-NPD performance.
H4: Internal perspectives influence S-NPD of the firm.
H5: Effective handling of external issues motivates S-NPD.
H6: Product life cycle analysis adds an additional feature to S-NPD on which the development of new products depends.
H7: Additional performance positively encourages S-NPD activities of the firm.

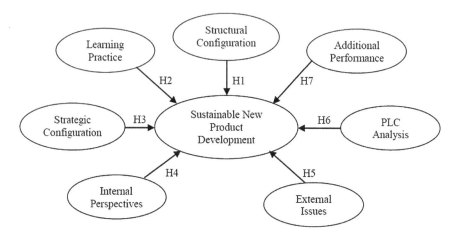

**Fig. 1** Path model of latent constructs with their estimated hypotheses

## 2.3 Development of Questionnaire

A semi-structure questionnaire is developed for accumulating the primary data from the industry experts. The questionnaire is divided into three sections. First section gathers the information regarding the respondent's profile. The second section comprises input CSFs and their manifest variables collecting the information about the input variables. The last and final section acquires manifests of output construct which is S-NPD. The second section is again segmented into two subsections. One is degree of importance of the manifest variables, and another is their rate of implementation in practical field. Seven-point Likert scale has been used to quantify the response of the samples. In this scale, 1 represents strongly agree and 7 strongly disagree for recording the importance of the manifests. In case of implementation and output, 1 shows very low and 7 very high implementation rate. A scope for sharing the own views of respondents for additional manifest variables is also provided for further value addition to the questionnaire.

## 2.4 Sample and Data Collection

This empirical research is performed considering the scenario of Indian manufacturing industries. Experts from small and medium-scale manufacturing companies mainly developing the engineering products are considered for this analysis. A pilot study based on the developed questionnaire is conducted by surveying from 36 experts of Kolkata and Howrah mainly for the content validation of the developed questionnaire. Design and development experts are treated as the targeted samples for the survey through direct interviewing or telephonic interview. Demographic profiles of the respondents are mentioned in Table 1.

## 3 Results

### 3.1 Analysis of Measurement Validity

Principal component-based factor analysis is performed to test the loadings of manifest variables for dimension reduction purpose. Variables having loadings less than 0.6 are rejected as per the conventional practice. In Table 2, the values of factor loadings are enlisted showing that all of the 34 variables have the loading values greater than 0.60 such as 0.676 to 0.675. This implies all of the manifests are considered for the framework development. The values of CR, $\alpha$ and AVE are also mentioned in Table 2 as obtained from the analysis of IBM SPSS AMOS 21.0 software. The values of CR range from 0.75 to 0.84, and values of $\alpha$ range from 0.720 to 0.858. In case of AVE, the values are from 0.46 to 0.59. It shows that values of reliability

**Table 1** Number of respondents from various manufacturing sectors of India

| Sectors | Number of respondents | Sectors | Number of respondents |
|---|---|---|---|
| Fabrication | 46 | Hydraulics and pneumatics | 25 |
| Electrical equipments | 33 | Burner and heater | 22 |
| Industrial valves | 32 | Material handling equipment | 21 |
| Textile machineries | 27 | Cell and battery | 14 |
| Firefighting equipment's | 26 | R&D sectors | 9 |
| Total respondents = 255 | | | |

indices (CR and $\alpha$) and discriminant validity (AVE) are greater than their threshold values depicting the reliability of the collected data. In case of $\alpha$, there are few values slightly less than 0.8, but they are also considered as reliable as their values are greater than 0.7.

## 3.2 Measurement Model Results

CFA is performed for estimating the unidimensionality of the model fit. The SRWs of the manifest variables are calculated showing the weights of the manifests associated with their respective construct. The values of SRWs range from 0.23 to 0.99 (Table 2) showing the positive correlation. The validation of the measurement model is checked by estimating the model fitness. The fitness tests show that the model has good model-to-data fit as per the obtained values $\chi^2$/degrees of freedom $= 1.96$, RMSEA $= 0.053$, GFI $= 0.871$, AGFI $= 0.832$ (Chen 2016; Hu and Bentler 1998) with respect to the desired range of $\chi^2$/degrees of freedom $\geq 2$, RMSEA $= 0.05$ for good fit and 0.08 for moderate fit, GFI $= 0.90$ and AGFI $= 0.90$ (Byrne 2010).

## 3.3 Structural Model Results

Analysis of measurement model is followed by the structural model analysis. It represents the linkages among the latent constructs. The path estimates between the latent variables are calculated using maximum likelihood estimation. It shows that path values range from 0.29 to 0.69 as listed in Table 3. All the estimated values are positive which depict that relation between the constructs is positive which means

**Table 2** List of manifest variables of latent constructs including results of reliability testing

| Latent with their manifest variables including reliability indices | FL | SRWS |
|---|---|---|
| **Structural configuration [$\alpha = 0.858$; CR = 0.84; AVE = 0.59]** | – | – |
| 1. Organizational structure for supporting S-NPD (m1) | 0.836 | 0.36 |
| 2. Managerial support (m2) | 0.804 | 0.92 |
| 3. Adoption of methods essential for S-NPD (m3) | 0.801 | 0.77 |
| 4. Supplier involvement for successful NPD (m4) | 0.789 | 0.87 |
| 5. Customer involvement (m5) | 0.761 | 0.59 |
| 6. Role of R&D for assuring lower impact of newly developed products on environment (m6) | 0.723 | 0.70 |
| 7. Risk managing for eliminating negative environmental impacts generated from newly developed products (m7) | 0.711 | 0.67 |
| **Learning practice [$\alpha = 0.775$; CR = 0.78; AVE = 0.48]** | – | – |
| 1. Internal and external training regarding environmental awareness (m8) | 0.890 | 0.83 |
| 2. Learning to cooperate with external stakeholders (m9) | 0.844 | 0.59 |
| 3. Technological advisory within the firm for environment-friendly new product development (m10) | 0.829 | 0.97 |
| **Strategic configuration [$\alpha = 0.834$; CR = 0.82; AVE = 0.56]** | – | – |
| 1. Managerial insights about strategic relevance of S-NPD | 0.725 | 0.90 |
| 2. Variations in strategies according to the changes occur in the market | 0.708 | 0.79 |
| 3. Strategies adopted for continuous improvement of S-NPD | 0.688 | 0.60 |
| **Internal perspective [$\alpha = 0.851$; CR = 0.83; AVE = 0.57]** | – | – |
| 1. Availability of both tangible and intangible assets essential for S-NPD (people, technology, knowledge) | 0.935 | 0.64 |
| 2. Support and motivation for innovative strategies encouraging for developing new products | 0.895 | 0.23 |
| 3. Synchronization among the internal teams | 0.812 | 0.73 |
| **External issues [$\alpha = 0.790$; CR = 0.80; AVE = 0.51]** | – | – |
| 1. Government rules and policies for promoting S-NPD | 0.850 | 0.54 |
| 2. Impartiality in regulations for both SMEs and large-scale industries | 0.763 | 0.87 |
| 3. Governmental support for developing sustainable products | 0.741 | 0.68 |
| **Product life cycle analysis [$\alpha = 0.798$; CR = 0.79; AVE = 0.52]** | – | – |
| 1. Avoid those raw materials having hazardous impact on environment | 0.852 | 0.72 |
| 2. Use of recycled materials having no metal emissions | 0.816 | 0.43 |
| 3. All scrap metals are recycled into pure fractions | 0.756 | 0.94 |
| 4. Optimized design of the product to be developed | 0.731 | 0.28 |
| 5. Recycling of all valuable components of the newly developed product after the end of the life cycle | 0.693 | 0.31 |
| **Additional performance [$\alpha = 0.720$; CR = 0.75; AVE = 0.46]** | – | – |
| 1. Brand image for reputation of the firm | 0.788 | 0.81 |

(continued)

**Table 2** (continued)

| Latent with their manifest variables including reliability indices | FL | SRWS |
|---|---|---|
| 2. Acceptability to the employees and customers | 0.732 | 0.68 |
| 3. Capability of the firm to learn the issues regarding sustainable new product development | 0.676 | 0.90 |
| **Sustainable new product development [α= 0.840; CR = 0.81; AVE = 0.52]** | – | – |
| 1. Reduced cost and increased profitability | 0.884 | 0.63 |
| 2. Resource efficiency | 0.860 | 0.75 |
| 3. Customer satisfaction | 0.801 | 0.99 |
| 4. Reduction of environmental pollution created by the product | 0.799 | 0.75 |
| 5. Health and safety aspects | 0.757 | 0.98 |
| 6. Social aspects | 0.722 | 0.88 |
| 7. Life cycle analysis | 0.684 | 0.94 |

**Table 3** Statistics of path estimates depicting the linkage of latent constructs

| Path description | Hypothesis | Estimate | $t$ values |
|---|---|---|---|
| Structural configuration → S-NPD | H1 | 0.69 (***) | 11.221 |
| Learning practice → S-NPD | H2 | 0.29 (***) | 4.685 |
| Strategic configuration → S-NPD | H3 | 0.58 (***) | 9.674 |
| Internal issues → S-NPD | H4 | 0.41 (***) | 6.335 |
| External perspectives → S-NPD | H5 | 0.43 (***) | 6.940 |
| PLC analysis → S-NPD | H6 | 0.36 (***) | 5.800 |
| Additional performance → S-NPD | H7 | 0.54 (***) | 9.201 |

[***$p$ < 0.01]

proposed hypotheses are supported. The validation of the structural model is tested with $\chi^2$/degrees of freedom, RMSEA, GFI and AGFI. The values obtained show the good model-to-data fit as well ($\chi^2$/degrees of freedom = 1.35, RMSEA = 0.048, GFI = 0.893, AGFI = 0.862) (Chen 2016; Hu and Bentler 1998). The structured model comprising of both measurement model and structural model is represented in Fig. 2.

As shown in Fig. 2, it states that the CSFs of S-NPD, namely structural configuration, learning practice, strategic configuration, internal issues, external perspectives, PLC analysis and additional performance, are positively correlated with S-NPD. This depicts that the developed hypotheses are supported. The values of path estimates are enlisted in Table 3.

The proposed model comprises the path estimates between latent constructs ranging from 0.29 to 0.69. The corresponding $t$ values of the respected path linkage between the latent constructs are also mentioned. Based on these t values, the p value is obtained which shows the path estimates are significant for $p < 0.01$. This infers

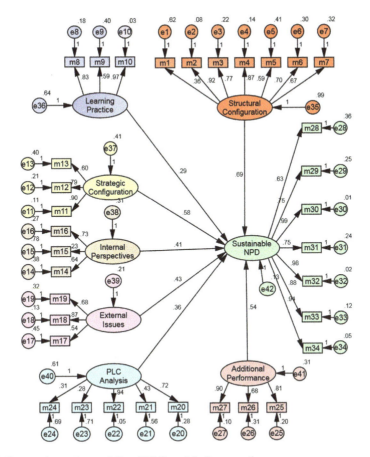

**Fig. 2** Structural equation modeling (SEM) model after execution

that the identified CSFs are all positively linked with S-NPD of the firm and they have significant impact on S-NPD in turn the success of the SMEs.

## 4  Discussion and Conclusion

The study realizes the factors of S-NPD and its impact on social, economic and environmental aspects. According to the analysis, it has been observed that structural configuration is the highest impacted factor succeeded by strategic configuration, additional performance, external perspectives, internal issues, PLC analysis and learning practice. From this study, the important role of managerial support to establish better structural configuration for S-NPD has been identified. It also depicts the vital role of technological advisory within the firm for environment-friendly new product development for enriching the learning practice. Similarly, for strategic configuration managerial insights about strategic relevance of S-NPD are the imperative most measured variable which must be taken care of. For internal issues and external perspectives, synchronization among the internal teams and impartiality in regulations for both SMEs and large-scale industries has the highest impact on their respective latent construct, respectively. Recycling of all scrap metals has the highest priority for PLC analysis, whereas capability of the firm to learn the issues regarding S-NPD is focused for better additional performance. In this study, S-NPD is treated as output construct which has been measured by social, economic and environmental attributes. Among these, customer satisfaction is the first priority of the firm followed by health and safety aspects, life cycle analysis, social aspects, resource efficiency and reduction of environmental pollution created by the product and reduced cost and increased profitability. Among these, resource efficiency and reduction of environmental pollution created by the product have the equal contribution to measure the success of S-NPD. Adoption of these practices not only motivates the successful implementation of the CSFs, but also ensures successful development of sustainable new products. Moreover, the Indian government must be aware of the environmental hazards and take hard steps by implementing rules and regulations to reduce the pollution created by the developed products. Scope of recycling in India is very much limited which must be taken care of for further development.

**Acknowledgements** The research work was substantially supported by a grant from the Department of Science and Technology (DST) of India as a DST INSPIRE Fellowship.

# References

Ackermann, F., & Eden, C. (2011). Strategic management of stakeholders: Theory and practice. *Long Range Planning, 44,* 179–196.

Aykol, B., & Leonidou, C. L. (2014). Researching the green practices of smaller service firms: A theoretical, methodological, and empirical assessment. *Journal of Small Business Management, 53*(4), 1264–1288.

Bertoni, A., Bertoni, M., Panarotto, M., Johansson, C., & Larsson, T. (2015). Expanding value driven design to meet lean product service development. *Procedia CIRP, 30,* 197–202.

Blackburn, W. R. (2007). *The sustainability handbook: The complete management guide to achieving social, economic and environmental responsibility.* Washington, D.C.: ELI Press. ISBN 978-1-1365-5202-1.

Blackburn, W. R. (2008). The sustainability handbook: The complete management guide to achieving social, economic and environmental responsibility (1st ed.). Environmental Law Institute. ISBN-10: 1585761745.

Boly, V., Morel, L., N'Doli, G. A., & Camargo, M. (2014). Evaluating innovative processes in French firms: Methodological proposition for firm innovation capacity evaluation. *Research Policy, 43,* 608–622.

Booz, A., & Hamilton. (1982). *New product management for the 1980's.* New York: Booz, Allen & Hamilton, Inc.

Byrne, B. M. (2010). *Structural equation modeling with amos: Basic concepts, applications, and programming.* New York: Taylor and Francis Group LLC.

Chen, H. C. (2016). The impact of children's physical fitness on peer relations and self-esteem in school settings. *Child Indicators Research, 9*(2), 565–580.

de Jesus Pacheco, D. A., Carla, S., Jung, C. F., Ribeiro, J. L. D., Navas, H. V. G., & Cruz-Machado, V. A. (2017). Eco-innovation determinants in manufacturing SMEs: Systematic review and research directions. *Journal of Cleaner Production, 142,* 2277–2287.

Ernst, H. (2002). Success factors of new product development: A review of the empirical literature. *International Journal of Management Reviews, 4*(1), 1–40.

Hansen, E. G., Grosse-Dunker, F., & Reichwald, R. (2009). Sustainability innovation cube—A framework to evaluate sustainability-oriented innovations. *International Journal of Innovation Management, 13*(4), 683–713.

Hillary, R. (Ed.). (2000). *Small and medium-sized enterprises and the environment.* Sheffield: Greenleaf Publishing.

Hu, L., & Bentler, P. M. (1998). Fit indices in covariance structure modeling: sensitivity to under parameterized model misspecification. *Psychological Methods, 3*(4), 424–453.

Ong, C. S., Lai, J. Y., & Wang, Y. S. (2004). Factors affecting engineers' acceptance of asynchronous e-learning systems in high-tech companies. *Information & Management, 41*(6), 795–804.

Paech, N. (2007). Directional certainty in sustainability-oriented innovation management. In M. Lehmann-Waffenschmidt (Ed.), *Innovations towards sustainability: Conditions and consequences* (pp. 121–140). Heidelberg: Physica. ISBN 978-3-7908-1650-1.

Paramanathan, S., Farrukh, C., Phaal, R., & Probert, D. (2004). Implementing industrial sustainability: The research issues in technology management. *R&D Management, 34*(5), 527–537.

Rennings, K. (2000). Redefining innovation e eco-innovation research and the contribution from ecological economics. *Ecological Economics, 32,* 319–332.

Schaltegger, S. (2011). Sustainability as a driver for corporate economic success: Consequences for the development of sustainability management control. *Society and Economy, 33*(1), 15–28.

Schaltegger, S., & Wagner, M. (2011). Sustainable entrepreneurship and sustainability innovation: Categories and interactions. *Business Strategy and the Environment, 20*(4), 222–237.

Schreiber, J. B., Nora, A., Stage, F. K., Barlow, E. A., & King, J. (2006). Reporting structural equation modeling and confirmatory factor analysis results: A review. *The Journal of Educational Research, 99*(6), 323–338.

Ullman, J. B. (2001). *Structural equation modeling.* In B. G. Tabachnick, & L. S. Fidell (Eds.), *Using multivariate statistics* (4th ed.). Needham Heights, MA: Allyn & Bacon. ISBN 0321189000.

# Clean Development Mechanism and Green Economy

**Md. Touseef Ahamad and M. Deepthi**

**Abstract** This paper delivers a brief description of Clean Development Mechanism and Green Economy and how CDM and Green Economy have its impact on sustainable development. We analyze here types of CDM Projects, steps for implementing a CDM project, the nature of CDM Projects in India which contributes India's developmental activities. It also includes various set of actions to achieve Green Economy in India. The objective of this paper is to assess the relationship between CDM Projects and Green Economy and its contribution to sustainable development. Green Economy is a development strategy to define economic and sustainable development. India occupies 2nd place in adopting CDM Projects.

**Keywords** Sustainable development · CDM Projects · Green Economy · Green jobs · International Society of Waste Management · Air and water

## 1 Introduction

About Clean Development Mechanism (CDM) & Green Economy. Under the Article 12, of Kyoto Protocol (Agenda21 1992), a program is introduced which was called Clean Development Mechanism CDM). Its objective is to reduce the greenhouse gases and prevents the ozone layer depletion. Developed as well as developing nations come forward to implement carbon emission reduction projects and policies. United Nations Climate Change Secretariat has analyzed respect of CDM Projects and stated the level and types of benefits. Expanding on the study in 2011, there are 4000 registered CDM Projects excluding programs of activities according to four topics, sustainable development, technology transfer, finance and regional distribution (Fig. 1).

It allows majorly developing countries to gain benefits from climate-friendly projects. India is second only to China in using CDM which helps to reduce its carbon emissions in India, and clean development projects from 2003 to 2011, a total of 2295

Md. Touseef Ahamad (✉) · M. Deepthi
Department of Mechanical Engineering, ANU, College of Engineering, Guntur, India
e-mail: touseefmd.anu@gmail.com

© Springer Nature Singapore Pte Ltd. 2020
S. K. Ghosh (ed.), *Waste Management as Economic Industry Towards Circular Economy*,
https://10.1007/978-981-15-1620-7_5_5

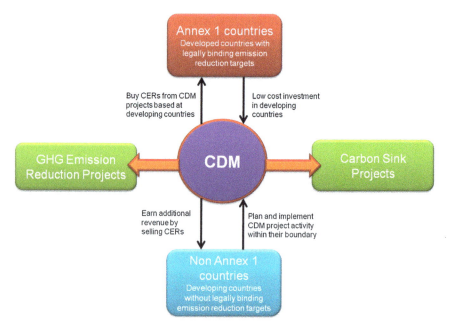

**Fig. 1** CDM

projects, around one-quarter of the global total had been registered with India's designated National Authority for the CDM (Agenda21 1992). The governing approach to CDM is best characterized as a "laissez-faire" system whereby the Indian Government neither actively promotes nor discourages CDM Project implementation in different states. The data analysis states that CDM Projects are concentrated in it more industrialized states such as Gujarat and Maharashtra in contrast; less industrialized states generally implement fewer CDM Projects (Alexeew et al. 2010). Fixed industrial capital plays an important role in CDM Project activity. India fixed industrial capital worth around 80 billion rupees. If the base of industrial capital increased to around 850 billion rupees, the state would implement around seven additional CDM Projects every year. This is mainly achieved by limiting the use of excessive available resources. Green Economy provides large areas of opportunities (Fig. 2).

Sustainable development became the overarching goal of the international community since in Conference on Environment and Development in 1992 (Alexeew et al. 2010). The concept of Green Economy has become a center of policy debates in recent years. A Green Economy is one whose growth in income and employment is driven by both public and private investments that reduce carbon emissions and pollution. Enhance energy and resource efficiency, and prevent the loss of biodiversity and ecosystem. The United Nations Environment Program [UNEP] launched an initiative in date-2008 called the Green Economy Initiative (GEI). Simply, we can say that the Green Economy is the aggregate of all activities operating with the primary intention of minimizing all forms of environmental impacts, climate changes,

**Fig. 2** Expected average
annual CERs by host party

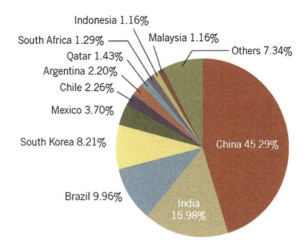

resource inefficiency, lack of fossil fuels, green fuels, etc. by adopting CDM Projects and Green Economy we have a lot of benefits if we manage it in a proper manner.

## 2  Literature Review

Data for the analysis was collected from publicly available documents UNFCCC [United Nations Frame Work Convention on Climatic Change]; UNEP [United Nations Environment Program], here I present reviews of some of the available literatures which give the glimpses of the CDM Projects and the concept of Green Economy and its various aspects (Alexeew et al. 2010).

## 3  Methodology

**Research Problem**
How a CDM Project contributes to sustainable development and what can be done to achieve Green Economy in India (Alexeew et al. 2010).

**Results and Discussions**

**Objectives**:
To achieve sustainable waste management in India through CDM and Green Economy and their impact.

**Objectives of Clean Development Mechanism are**

- Help slow and prevent climate change.
- Guide developing nations to develop sustainable methods.
- Encourage and help countries to innovate various ways to prevent environmental disorders.

**CDM Project types**

India, being a developing country, offers a wide range of potential CDM Project types out of 23 proposed CDM Projects on February 2004. Biomass contributes a large proportion of 51%, and other ones are HFC—23 (4%), biofuel (8%), municipal solid waste (4%), energy efficiency waste heat recovery (25%), fuel switch (8%) (Fig. 3).

**Steps for Implementing a CDM Project**

1. **Project Identification**:
   It is the first step where research is done.
2. **Government Endorsement**:
   After that, our idea should be presented before the Ministry of Environment, Forest and Climate Change to be endorsed by the Government of India.
3. **Project Development**:
   A study is done to measure the baseline as per the Kyoto Protocol and Marrakesh accord.

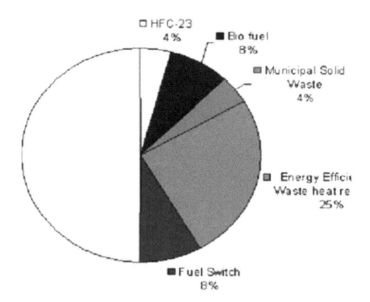

**Fig. 3** CDM Project types

4. **Validation**:

The result of the baseline identification survey is verified by the CDM Executive Body.

5. **Registration**:

A formal acceptance by executive body makes the identified project a CDM Project.

6. **Monitoring**:

The change is monitored from time to time.

7. **Verification**:

The data is verified and then sent for certification.

8. **Certification**:

This is the final step where the monitoring body certifies after proper verification (Fig. 4).

### CDM Project Co-benefits in Andhra Pradesh—India

**Aim**: To provide energy-efficient lighting for low-income households.

**Project Description**: Visakhapatnam OSRAM CFL Project involves the distribution of 7 lakhs long life compact fluorescent lamps (CFL) to households in the districts of Visakhapatnam, Andhra Pradesh, India (Government of India 2012).

**Benefits of the Project**: Reduction in poverty, access to energy-efficient lighting and empowerment of people status. Over 3000 women have been engaged through self-help groups for efficient lighting to low-income households. It also includes economic, social and environmental benefits.

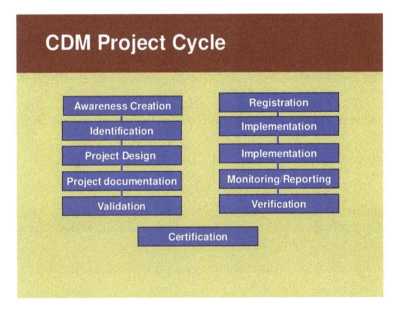

**Fig. 4** CDM Project cycle

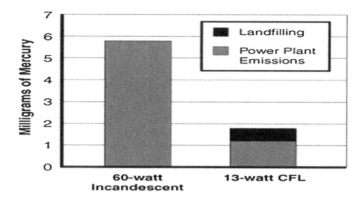

**Fig. 5** Total mercury emissions for CFL* and incandescent bulbs

**Project Facts**: Project title and numbers: Visakhapatnam OSRAM CFL Distribution Project

Project type and methodology: Energy efficiency in households—lighting, MS-11—demand side
Energy-efficient equipment
Location: Visakhapatnam, AP, India Lat 17°40'31°N
Long: 83°13'57 F (Fig. 5)
CFL*—compact fluorescent light
History and CERS: Registered February 12, 2009. Project operational life, 10 years. Expected total
CER's 274,270[tco$_2$ eq] CER's issued to date; request yet to be submitted.
Project Link: (https://cdm.unfcc.int/projects/DB/tuev_sued1206629154.85/view).

This factsheet is one of a series produced by the UNFCCC Secretariat to highlight the types of co-benefits generated by the CDM. This is one of the projects of CDM introduced in Andhra Pradesh, India.

## Some of the projects of CDM are their co-benefits (http://cdm.unfcc.int)

(1)  Low-cost energy services to rural households in Kolar District, India [CDM Project 121].

### Key benefits
Reduction in poverty, access to clean energy for cooking and empowerment of the community.

(2)  Community-led approaches to waste management in Bali, Indonesia.

### Key benefits
Empowering communities to take action on waste management [CDM Project 1885].

(3)  Effective use of firewood in rural communities in Guinea Savannah Region, Nigeria [CDM Project 2711].

**Key benefits**
Providing access to energy-efficient cooking.

(4)  Micro-hydro-power contributing to sustainable rural electrification in Chende-bji, Bhutan.

**Key benefits**
Providing electricity to, and improving the livelihoods of, isolated mountain communities [CDM Project 62].

(5)  Enhancing living conditions through energy upgrades to low-cost housing in Capetown, South Africa [CDM Project 79].

**Key benefits**
Reducing cost, improving living condition and generating jobs and skills.

(6)  Micro-hydro-power delivering community infrastructure and services in the Sayan District, Lima, Peru [CDM Project 88].

**Key benefits**
Investing CDM revenues back into local community projects.

(7)  Rapid and reliable bus transport for urban communities in Bogota, Colombia [CDM Project 672].

**Key benefits**
Improving the quality of life for urban communities.

(8)  Biogas digesters enhance the welfare of low-income rural communities in Hubei province, China [CDM Project 2221].

**Key benefits**
Improving health and welfare creating employment and enhancing incomes in rural communities.

There are a lot of CDM Projects, but few are mentioned in this paper. The CDM is one of the mechanisms by which they could be transferred. The Intergovernmental Panel on Climate Change [IPCC] defines technology transfer as "a broad set of processes and equipment for mitigating and adapting to climate change amongst different stake holders".

## Recent results about technology transfer through CDM states that

- The frequency of technology transfer differs significantly by project type.
- The host countries are greatly involved in technology transfer.
- It falls as the number of projects of the same type in a host country increases.

These CDM Projects also depends upon transaction costs, project risks, development dynamics and climatic change. The performance of CDM Projects also depends on Green Economy. My research includes Green Economy, Environmental and Health Impact.

The Green Economy is the aggregate activity whose primary purpose is reducing all forms of environmental impact, enhance energy and resource efficiency and prevent the loss of biodiversity and ecosystems (Barbier 2011).

The Green Economy is defined by the United Nations Environment Program [UNEP] as one that results in "improved human well-being and social equity, while significantly dipping the environmental risks & ecological scarcities".

The Nagoya Protocol takes the leading step to protect green governance in a multilateral environmental agreement by recognizing four key rights of communities.

- Traditional knowledge related to genetic resources.
- Self-governance through customary laws and community protocols.
- Benefit from the utilization of their traditional knowledge and genetic.

## These are actions which can be done to achieve Green Economy

- **Buildings**: Energy audit can reduce the building's foot print that leads to significant savings in energy costs (www.energystar.gov).
- **Fisheries**: We can avoid overfishing by working to promote sustainable practices.
- **Forestry**: Deforestation is the main cause of greenhouse gas emissions. We can find alternative way to meet our demands by using technology.
- **Agriculture**: We can follow sustainable agriculture practices these by we can feed every one with healthy food. In this way we support Green Economy for agriculture (Asian Development Bank 1998). Agriculture is facing a horde of challenges like demand-side challenges [food security, population growth, growing pressure from biofuels] and supply-side challenges [limited availability of land, water and mineral inputs]. We can meet these challenges by increasing farm productivity, rebuilding the ecological resources, improving water management by enabling conditions such as governance, regulations and taxes and trade laws and ag subsidies.

- **Waste**: Recycling appropriate materials and composting food waste reduces the demand on our natural resources.

A green organization is well-defined as one that produces goods and services to minimize environmental impact. Green skills are the knowledge, training or experience as they narrate to technologies or materials that diminish environmental impact (Ministry of Environment and Forest 2004).

**The top four areas of opportunity within the Green Economy**

- Renewable energy and energy efficiency
- Buildings, retro-fitting and construction
- Transportation and alternative transportation and
- Waste recycling and waste management (Fig. 6).

The two major forces which may be forcing companies to adopt green practices are government regulation and public acceptance of products and public attitudes (The future of sustainability: Rethinking environment and development in the Twenty first century).

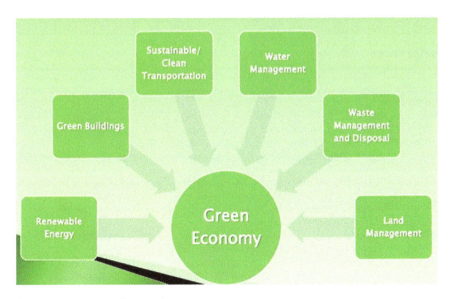

**Fig. 6** Green Economy framework

# 4    Conclusion

We know that the country's development policies aimed at reducing the widespread poverty, regional imbalances and enhancing the quality of people's lives remain the key focus of India's policy agenda. CDM offers the possibility of developing projects which lead to sustainable development the CDM Projects that are currently proposed in India which are therefore all small in size. Hopefully, larger projects may be prompted after initial learning process near completion. One of the key challenges that need addressing is to demystify the inherent complexities of the mechanism resting on a counterfactual foundation. The CDM was plugged into the Kyoto Protocol without it having gone through intensive prior discussion. A key challenge for negotiators for the period beyond Kyoto is how to align climate actions in developing countries with their development priorities CDM which will remain the principle vehicle for cooperative actions between developed and developing countries. The CDM and Green Economy are interrelated concepts case which has to be taken to ensure that the "Green Economy" term and concept is also understood to include the social, equity and development dimensions including the need for the international provision of finance and technology. Global economic reforms and the risks of the misuse of the term are adequately addressed.

# References

Agenda21. (1992). *United nations conference on environmental and development*, Rio de Janeiro, Brazil, June 3–14, 1992.

Alexeew, J., Bergset, L., Meyer, K., Petersen, J., Schneider, L., & Unger, C. (2010). An analysis of the relationship between the additionality of CDM projects and their contribution to sustainable development. *International Environmental Agreements, 10*(3), 233–248.

Asian Development Bank (ADB). (1998). *Asia least-cost green house gasses abatement strategy (ALGAS)*. Manila: ADB.

Barbier, E. (2011). *The policy challenges for green economy and sustainable economic development*. Natural resources Forum 35, United Nations.

Government of India. (2012). *Interim report on the expect group on low carbon strategies for inclusive growth*. Planning Commission.

Ministry of Environment and Forest. (2004). *India's National Communication of the United National Frame work convention on climate change, executive summary*.

The future of sustainability: Rethinking environment and development in the Twenty first century. *Report of the international union for conservation of Nature*. Renowned thinkers meeting, 29–31, January 26.

http://cdm.unfcc.int.

www.energystar.gov.

# A Review on Biodegradable Packaging Materials in Extending the Shelf Life and Quality of Fresh Fruits and Vegetables

M. Ramesh, G. Narendra and S. Sasikanth

**Abstract** Different packaging techniques were employed to reduce the post-harvest losses nearly about 30–40% of total production of fruits and vegetables, to improve the quality and safety aspects of the produce, and to combat the market price which in turn increases demand in the market to earn profits. In order to maintain the quality and safety aspects of fruits and vegetables, many traditional technologies were adapted but that does not suffice the requirement in maintaining the extended shelf life. In the recent past, several emerging technologies which are eco-friendly have a huge potential which was used to maintain the safety and quality parameters of fruits and vegetables but still has certain limitations for each of its techniques.

**Keywords** Packaging · Nanocomposite · Edible films · Biodegradable · Shelf life · Eco-friendly

## 1 Introduction

### 1.1 Modified Atmospheric Packaging

One of the most popular practices used for extending the shelf life of fruits and vegetables, by combination of technologies i.e., using controlled atmosphere (depleting the oxygen ($O_2$) and to increase carbon dioxide ($CO_2$) levels) and refrigeration to increase the freshness which helps to meet the opportunity more for consumer consumption (Dilley 2006) Table 1.

Another best practice, to increase the shelf life is by Controlled Atmosphere (CA) storage in this optimum concentration of different levels of gas consumption storage regimes such as Ultra Low Oxygen (ULO), Low Oxygen (LO), Dynamic Controlled Systems (DCS)/Atmosphere (DCA) were maintained and finds its exclusive benefits over combination of technologies with controlled atmosphere and refrigeration. CA

M. Ramesh (✉) · G. Narendra · S. Sasikanth
Institute of Science and Technology, Jawaharlal Nehru Technological University, Kakinada, AP, India
e-mail: ramesh_biotech@jntuk.edu.in

© Springer Nature Singapore Pte Ltd. 2020
S. K. Ghosh (ed.), *Waste Management as Economic Industry Towards Circular Economy*,
https://doi.org/10.1007/978-981-15-1620-7_5_6

**Table 1** Application of different concentrations of oxygen and carbon dioxide gases for various commodities of fruits and vegetables

| Oxygen ($O_2$) %[a] | Carbon dioxide ($CO_2$) %[a] | Commodity |
|---|---|---|
| *Fruits* | | |
| 2–3 | 2–3 | Apricot |
| 2–3 | 5–10 | Avocado |
| 2–5 | 3–5 | Banana |
| 5–10 | 15–20 | Blackberry |
| 5–10 | 15–20 | Blueberry |
| 3–10 | 10–15 | Cherry |
| 1–2 | 0–5 | Cranberry |
| 5–10 | 15–20 | Fig |
| 2–5 | 1–3 | Grape |
| 3–10 | 5–10 | Grapefruit |
| 1–2 | 3–5 | Kiwi |
| 3–5 | 5–10 | Mango |
| 1–2 | 3–5 | Nectarine |
| 5–10 | 0–5 | Orange |
| 3–5 | 5–10 | Papaya |
| 1–2 | 3–5 | Peach |
| 3–5 | 5–8 | Persimmon |
| 3–5 | 5–10 | Pineapple |
| 1–2 | 0–5 | Plum |
| 5–10 | 15–20 | Raspberry |
| 5–10 | 15–20 | Strawberry |
| *Vegetables* | | |
| 21 | 10–14 | Asparagus |
| 2–3 | 2–3 | Artichokes |
| 2–3 | 3–7 | Beans |
| 1–2 | 5–10 | Broccoli |
| 1–3 | 5–10 | Brussels sprouts |
| 2–3 | 4–6 | Cabbage |
| 2–3 | 3–4 | Cauliflower |
| 2–4 | 3–5 | Celery |
| 3–4 | 4–5 | Chicory |
| 1–2 | 5–10 | Leeks |
| 1–3 | 0 | Lettuce |
| 21 | 10–14 | Mushrooms |
| 21 | 4–10 | Okra |

(continued)

**Table 1** (continued)

| Oxygen ($O_2$) %[a] | Carbon dioxide ($CO_2$) %[a] | Commodity |
|---|---|---|
| 9–10 | 8–10 | Parsley |
| 1–2 | 2–3 | Radish |
| 3–5 | 2–3 | Tomato |

[a] A mole percentage or volume; the remaining occupied by nitrogen gas (Table 3)

**Table 2** List of vendors that produce active and intelligent packaging systems

| Company | Country | Web page |
|---|---|---|
| BASF SE | Germany | www.basf.com |
| Amcor | Australia | www.amcor.com |
| Honeywell International Inc. | United States | www.honeywell.com |
| Landec Corporation | United States | www.landec.com |
| Bemis Company | United States | www.bemis.com |
| Crown Holdings Inc. | United States | www.crowncork.com |

storage for fresh cut fruits and vegetables is useful in preventing the diseases and to reduce the storage disorders (Prange et al. 2005). In order to retain the firmness and to maintain the quality of fruits, hydro- or forced-air cooling is used to reduce the $O_2$ levels by purging the nitrogen which in turn considered as rapid controlled atmosphere storage for depleting the ethylene synthesis (Vigneault and Artes-Hernandez 2007). Various limitation to implant this newer technologies of CA was due to its low financial returns and adaption of this technologies to high value commodities storage of fruits and vegetables.

## 1.2  Active Packaging and Intelligent Packaging

Active food packaging is an innovative approach keeping in view the consumer demands like more convenience, shelf life improvement which includes the study of various processes like physical, chemical and microbiological aspects of appropriate packaging systems. During transportation and in order to avoid cross-contamination for fresh fruits and vegetables, various such challenges need to be taken into account to increase the shelf life, quality and food safety and to improve the sensory properties by modifying the conditions in the packaging systems (Vermeiren et al. 1999). Intelligent packaging helps to monitor the packaging conditions such as tracing the carbon dioxide and oxygen levels, detecting with respect to time vs temperature history, recording/identifying the foodborne pathogens and sensing mechanical, chemical and enzymatic reactions (Wilson 2007). Commercial Active and Intelligent packaging systems from various vendors are given in the Table 2.

Modified atmospheric packaging technique can be used to alter the gases mixture composition, thus decreasing the oxygen levels and maintaining the carbon dioxide levels depending upon the type of fruits and vegetables, but this technology has its own limitations with respect to lipid oxidation (chemical process), microbial growth spoilage (microbiological studies) fruits and vegetables respiration (physiological process). All these various processes can best be controlled by different ways using active packaging systems and thus increases the shelf life of packaged produce (Yam 2010).

Techniques like moisture scavenging, barrier packaging and releasing systems are predominating the market of around 80–85% rather than conventional packaging systems (Robinson and Morrison 2010).

Combination of biodegradable with biopolymer packaging films like use of whey protein and ascorbic acid helps in regulating the oxygen levels and acts as a scavenger for oxygen along with improvement in the mechanical as well as barrier properties which make the way for commercial applications (Prommakool et al. 2011).

Depending upon the moisture content of foods such as for minimally processed fruits and vegetables and marine products, the type of technique is employed for emission of carbon dioxide by reaction of acidulates like acetic acid with hydrating agent like water and sodium bicarbonate in the presence of utilization of moisture in the packaged foods in order to activate the emission process. This type of process is not applicable for the low moist foods (Ozdemir and Floros 2004). More recently for ethylene absorbers, scrubber was developed to continuously remove the ethylene from the environment (Martinez Romero et al. 2009).

The primary cause for deterioration of fruits and vegetables is due to growth of microorganisms on the surface; to prevent this spoilage, various antimicrobial compounds like silver, zeolite, chitosan, bacteriocins, some organic acids, etc., which possess antimicrobial activity are used in packaging systems.

## 1.3   Edible Packaging Films

Most of the edible coatings for packaging films was used as a primary packaging material for fresh cut fruits and vegetables this enables to act as a barrier against oxygen, moisture and carbon dioxide thus protecting from spoilage. Additionally advantages are follows i.e., by lowering the gaseous exchange, reducing physiological disorders, respiration rate is reduced, texture improvement and volatile compounds were thus retained (Rojas-grau et al. 2009).

By incorporating micronutrients like vitamins, essential minerals and essential fatty acid components into the edible films will help to improve the nutritional properties of fruits and vegetables which are less in micronutrients.

## 2    NanoComposites for Fruits and Vegetable Packaging

Currently, the use of nanoparticles in commercial packaging systems has less application for fruits and vegetable packaging. Nanoparticles in the range of 10–100 nm are widely used for exhibiting the reactivity of the material by any of the two approaches either self-assembly or bottom-down and top-down; in these processes, surface to volume ratio is increased and thus helps to increase the surface particles number (Ward and Dutta 2005).

Most of the active packaging and conventional-type packaging are replaced by nanocomposite films to increase the quality, safety and shelf life aspects of the packaged food products. In order to have a uniform distribution of antimicrobial activity, the use of nanoclay with antimicrobial agent in the polymeric blend helps to improve the mechanical and antimicrobial properties of packaging systems. Thus, various nanoparticles were used such as aluminum hydroxide, carbon tubes, silicates and titanium oxides (Pandey et al. 2005).

Antimicrobial compounds are used in nanocomposite films directly in compounds form or as coating itself to control the growth of microbial spoilage (Appendini and Hotchkiss 2002; Persico et al. 2009). The use of silver nanoparticles as antimicrobial packaging material that releases active biocide substances into the system to improve the quality, safety and helps in delaying the spoilage of fresh cut fruits and vegetables. Many researchers stated the use of titanium oxide nanoparticles in the packaging shows greater antimicrobial properties than the silver nanoparticles (Duncan 2011). Concern regarding the use of nanoparticles stating that these are more toxic than their counterparts of macro-sized particles which do not show any biological activity as like that of micro-sized particles (Taylor 2008).

Due to consumer awareness to novel packaging techniques like active packaging, the use of nanoparticles has made a new era and widened up the research area especially in developing countries like India to invest more on the quality and safety aspects. More studies need to be conducted on the interaction of this emitters and absorbers on the packaging conditions. Also, studies should focus on the use of antimicrobial compounds used in food packaging systems, especially for fruits and vegetables. There is no doubt to say that the use of this novel packaging tends to improve the textural, mechanical and antimicrobial properties, but keeping in mind the cost of the package, the present packaging machineries in food industries which also need to be utilized efficiently are the challenges faced by this new packaging techniques implementation. The use of nanoparticles in active packaging systems need to be concentrated more on the permeability, releasing into the packaged foods and susceptibility of the packaged films.

The use of modified atmospheric packaging individually or in combinations with other techniques has increased the demand in recent trends because of its longer storage period due to various aspects like transportation, wholesaler and retailer, etc. Various combinations of gases like nitrogen, carbon dioxide and oxygen were used depending upon the type of produce along with use of scavengers within the package. Modified Atmospheric Packaging (MAP) in combination with intelligent packaging

**Table 3** List of vendors that produce modified atmospheric packaging

| Company | Country | Web page |
|---|---|---|
| Bemis | United States | www.bemis.com |
| Berry Plastics | United States | www.berryglobal.com |
| Dansensor | Denmark | https://Dansensor.com/ |
| Linde | Germany | www.linde.com |
| Sealed Air | United States | https://sealedair.com/ |
| Amcor | Australia | www.amcor.com |
| G. Mondini | Itlay | www.gmondini.com |
| LINPAC | England | www.linpacpackaging.com |
| ORICS | United States | www.orics.com |
| Point Five Packaging | Chicago | www.p5pkg.com |
| StePac | Brazil | www.stepac.com |
| Total Packaging Solutions | India | www.totalpackagingsolutions.in |
| Harpak-ULMA Packaging | United States | www.harpak-ulma.com |
| Winpak | Canada | www.winpak.com |

technique helps in monitoring the release of ethanol gas and also to detect the depletion of oxygen levels which will be useful to retain the freshness and odor/aroma of the package food (Zhang et al. 2007).

The use of individual modified atmosphere for fruits and vegetables have not been investigated in detailed. There is no clear understanding that use of MAP will enhance the physiological disorders, and mechanism for the initiation of these disorders are not depicted. Further investigation needs to be evaluated on the use of MAP with novel technologies like intelligent and active packaging techniques, and it is important to study the microbial pathogens and foodborne parasites on this produce. Modified Atmospheric Packaging (MAP) with vendors list (Table 3) from various countries and application of various concentrations of gases for different commodities were also given Table 1.

# 3   Conclusion

Biodegradable packaging materials have been the topic of many studies of last two decades. The current applicability of biodegradable packaging materials for both short shelf life as well as for long shelf life for fresh cut fruits and vegetables. Some of the obstacles in the development potential have to overcome, the improved water vapour permeability, cost effectiveness. Full commercial production and its application is still some way off, but the potential regarding biodegradable packaging materials is now being realised.

**Acknowledgements** I sincerely thank Science and Engineering Research Board (EEQ/2016/000031 DT: 17-01-2017, SERB, Govt, New Delhi-110070) for funding this entire project and also to JNTUK, Kakinada, for enabling me to continue my research project.

**Conflict of Interest** The authors declared that they have no conflict of interest.

# References

Appendini, P., & Hotchkiss, J. H. (2002). Review of antimicrobial food packaging. *Innovative Food Science and Emerging Technologies, 3*(2), 113–126.

Diley, D. R. (2006). Development of controlled atmosphere storage technologies. *Stewart Postharvest Review, 2*(6), 1–8.

Ducan, T. V. (2011). Applications of nanotechnology in food packaging and food safety barriers materials, antimicrobials and sensors. *Journal of Colloid and Interface Science, 363*(1), 1–24.

Martinez-Romero, D., Guillen, F., Castillo, S., Zapata, P. J., Valero, D., & Serrano, M. (2009). Effect of ethylene concentration on quality parameters of fresh tomatoes stored using carbon-heat hybrid ethylene scrubber. *Postharvest Biology and Technology, 51*(2), 206–211.

Ozdemir, M., & Floors, J. D. (2004). Active food packaging technologies. *Critical Reviews in Food Science and Nutrition, 44*(3), 185–193.

Pandey, J. K., Reddy, K. R., Kumar, A. P., & Singh, R. P. (2005). An overview on the Degradability of polymer nanocomposite. *Polymer Degradation and Stability, 88*(2), 234–250.

Persico, P., Ambrogi, V., Carfagna, C., Cerruti, P., Ferrocino, I., & Mauriello, G. (2009). Nanocomposite polymer films containing carvacrol for antimicrobial active packaging. *Polymer Engineering and Science, 49*(7), 1447–1455.

Prange, R. K., Delong, J. M., Daniels-Lake, B. J., & Harrison, P. A. (2005). Innovation in controlled atmosphere technology. *Stewart Postharvest Review, 1*(30), 1–11.

Prommakool, A., Sajjaanantakul, T., Janjarasskul, T., & Krochta, J. M. (2011). Whey protein polysaccharide fraction blend edible films: tensile properties, water vapour oxygen permeability. *Journal of the Science of Food and Agriculture, 91*(2), 362–369.

Robinson, D. K. R., & Morrison, M. J. (2010). Nanotechnologies for food packaging: Reporting the science and technology research trends. *Report for the Observatory Nano, 15,* 476–480.

Rojas-Grau, M. A., Soliva-Fortuny, R., & Martin-Belloso, O. (2009). Edible coating to incorporate active ingredients to fresh-cut fruits: A review. *Trends in food science & technology, 20*(10), 438–447.

Taylor, M. R. (2008). *Assuring the safety of nanomaterials in food packaging: The regulatory.*

Vigneault, C., & Artes-Hernandez, F. (2007). *Gas treatments for increasing the phytochemical content of fruits and vegetables.*

Vermein, L., Devlieghere, F., Van Beest, M., De Kruijf, N., & Debevere, J. (1999). Developments in the active packaging of foods. *Trends in Food Science & Technology, 10*(30), 77–86.

Wilson, C. L. (Ed.). (2007). *Intelligent and active packaging for fruits and vegetables.* CRC press.

Ward, H. C., & Dutta, J. (2005). *Nanotechnology for agriculture and food systems: A view.* Klong Luang: Micro elements, school of advanced technologies, Asian institute of technology.

Yam, K. L. (Ed.). (2010). *The Wiley encyclopedia of packaging technology.* Wiley.

Zhang, Y., Liu, Z., & Han, J. H. (2007). *Modeling modified atmosphere packaging for fruits and vegetables.* Intelligent and active packaging for fruits and vegetables, 165.

# Wealth from Poultry Waste

V. V. Lakshmi, D. Aruna Devi and K. P. Jhansi Rani

**Abstract** Poultry farming is practiced intensively throughout the world which generates huge quantities of nitrogen-rich waste in the form of poultry litter and feather waste. Feather which is made of almost pure keratin protein is generated in bulk quantities as a by-product of poultry industry all over the world. It is estimated that 400 million chickens are processed every week with huge amount of feather produced as waste globally. Though made of pure keratin protein, the by-product is neither profitable nor environment friendly. Keratin is highly recalcitrant to all common proteases being slowly digested/degraded in the environment leading to dumps thereby contributing to global environmental pollution problem. Keratin waste has not been considered as a source of dietary protein or organic manure (OM) till recently, as value of FM produced traditionally is very poor with locked nutrients thus not serving as good products. Organic farming has gained popularity due to high health risks associated with the use of chemical fertilizers. Organic produce is sold in the market at almost double price compared to those produced by using chemical fertilizers. Technology has been developed in SPMVV for efficient degradation of poultry waste in five days by developing native bacteria. Feather meal produced by keratinase treatment was found to be significantly superior in nutritive value compared to ones produced by traditional means thus increasing their economic value. KTF had higher value as compared to farmyard manure and vermicompost, the commonly used OM in terms of water retention capacity and production of KTF in shorter time at a lower cost. KTF can also be utilized as feed supplement in poultry and aquaculture industry. Digestion of keratin waste has high potential to serve as a cheap source for the production of value-added products having a high commercial value.

**Keyword** Chain

V. V. Lakshmi (✉) · D. Aruna Devi (✉) · K. P. Jhansi Rani
Department of Microbiology, Sri Padmavati Mahila Visvavidyalayam, Tirupati, India
e-mail: lakshmivedula28@gmail.com

© Springer Nature Singapore Pte Ltd. 2020                                          67
S. K. Ghosh (ed.), *Waste Management as Economic Industry Towards Circular Economy*,
https://10.1007/978-981-15-1620-7_5_7

# 1 Introduction

Keratin represents a major class of cellular-derived, fibrous, insoluble structural protein of animal origin. It is found in biological derivatives of ectoderm such as hair, wool, scales, feathers, quills, nails, hoofs, horns and silk found in all vertebrates. Hair and feather are examples of almost pure keratin which is highly recalcitrant to all common proteases like trypsin, pepsin and papain, which is due to its strong cross-linked rigid polypeptide chains leading to a very slow digestion/degradation in the environment (Brandelli et al. 2015; Bach et al. 2015). Elementally, feather is composed of 45% carbon, 14% nitrogen, 2.9 g/kg phosphorus, 1.5 g/kg potassium and 0.8 g/kg magnesium. Feather keratin has an elevated content of glycine, alanine, serine, cysteine and valine, but has lower amounts of the amino acids like lysine, methionine and tryptophan (Govern 2000). Significant amount of keratin waste is generated as by-products of agro-industrial processing. Poultry farming is practiced intensively throughout the world which generates huge quantities of nitrogen-rich waste in the form of poultry litter and feather waste (Onifade et al. 1998; Brandelli et al. 2015). It is estimated that 400 million chickens are processed every week and globally, approximately 8.5 million metric tonnes of poultry waste are generated annually. India ranks fifth in poultry industry and contributes about 3.5 million tons of this waste alone. In addition to the feather waste, poultry industry produces huge quantities of farm litter and animal waste from slaughter houses which is rich in keratin. These emit strong fouling smell and attract flies and ultimately find way into water bodies, thereby significantly contributing to environmental pollution (Jayathilakan et al. 2012). Keratin waste has not been considered seriously as a source of dietary protein, rich source of amino acids or organic manure (OM) till recently, as value of feather meal produced traditionally is very poor and at best barely covered its cost of production. Being slowly degraded in nature, the nutrients are locked hence do not serve as good manure also (Mortiz and Latshaw 2001; Grazziotin et al. 2006). Various strategies are adopted to handle the volume of this waste produced continuously. Traditionally, feather waste generated by the member firms is disposed off to waste disposal sites, landfills, incinerated. The traditional methods of feed production yield poor quality of feed with low commercial value due to destruction of heat-labile amino acids to formation of non-nutritive amino acids, thereby limiting the product value as feed supplements. Slow mineralization of feather in nature does not make it an effective fertilizer. Thus, almost pure keratin protein produced in huge quantities is neither profitable nor environmentally friendly forming a produce of high volume with low profit margin. Hence, bulk of the waste produced piles up as dumps, thus making the disposal of this waste a global environmental problem contributing to pollution (Papadopoulos 1984; Grazziotin et al. 2006; Brandelli et al. 2010).

Though, bioconversion has long been thought to be an attractive and viable option to utilize the keratin-rich waste as some of the microorganisms could degrade keratin by producing keratinase enzyme. Till 1990s, the production of keratinase was reported mainly from mesophilic fungi and actinomycetes. Strains of *Doratomyces microsporum*, *Aspergillus fumigatus* and *Aspergillus flavus* producing keratinases are also identified (Santos et al. 1996; Gradisar et al. 2000; Gupta and Ramnani 2006). These findings were mainly of academic interest as it required fairly long time (ranging up to 40 days) to bring ≥50% digestion of keratin and organisms being pathogens further limited their application potential. The growing prominence of keratinases in the last decades is due isolation of several non-pathogenic microorganisms, which degraded keratin by secreting a specific proteases keratinase. In 1990, Williams et al. isolated and characterized *B. licheniformis* isolate, which was able to degrade native feathers, and subsequently several other strains of *Bacillus licheniformis* and *Bacillus subtilis* have been identified which could hydrolyze both native and denatured keratin. (Gupta and Ramnani 2006; Suneetha and Lakshmi 2004; Jeevana lakshmi 2008; Jeevana Lakshmi et al. 2013). Thus, the focus of studies in the last two decades has shifted to isolating non-pathogenic microorganisms with good keratinolytic potential. Most of the keratinases reported have been identified to be serine proteases (Lin et al. 1992; Bockle et al. 1995; Friedrich and Antranikian 1996; Suh and Lee 2001), though a few metalloproteases and acidic keratinases are also reported (Allpress et al. 2002; Farag and Hassan 2004). Keratinases also have wide applications in a number of sectors like feed, fertilizer, detergent, leather, textile cosmetic and pharmaceutical industries and biomedical applications. In leather industry Keratinase is employed to improve the quality of leather produced as well as for tannery effluents treatment to significantly reduce toxicity of effluents before the release (Gupta and Ramnani 2006; Kumari et al. 2015a). The vast application potential for keratinase has resulted in a drive for production of keratinase by fermentation at industrial scale. Though highly promising, the full commercial potential of keratinases is yet to be realized. The biodegradation of agro-industrial waste like feather and its efficient recycling increases energy conservation and reduces environmental pollution load (Brandelli et al 2010). The major limiting factor in the wide scale usage of keratinases is mainly the availability of efficient and cost-effective method for production of keratinases in large scale. In view of the economic importance, microorganisms with keratinase activity were isolated from Tirupati and developed to produce high amounts of enzymes. The application potential of the keratin treated feather (KTF) as organic manure was evaluated which exhibited high nutritive value and enhanced water retention capacity of the soil.

## 2   Materials and Methods

Screening for keratinolytic organisms from soil samples (collected from poultry farms and poultry litter from Tirupati) led to the isolation of *B. subtilis* sp. Strain improvement and optimization of parameters of fermentation resulted in designing a cost-effective fermentation media with starch as a carbon source and soya bean meal as nitrogen source with a yield of >500 KU/ml. Semisolid state fermentation was developed using feather and agricultural waste like black gram husk (BG) and groundnut husk for biodegradation of feather using keratinase producing organisms (Kumari 2011; Kumari and Lakshmi 2015b). For each batch, 20 g ball milled feather,

20 g black gram husk and 20 g of groundnut husk were taken in 2 l conical flask, and 250 ml of mineral media (NaCl-0.5 g, $KH_2PO_4$-0.3 g, $K_2HPO_4$-0.4 g, $MgCl_2$-0.1 g per liter) was added. The media was sterilized at 10 lb/$inch^2$ for 15 min and inoculated with overnight culture 10 ml of *Bacillus* sp. BF 20 culture (~$10^9$ CFU/ml) producing keratinase enzyme. The flasks were incubated at 37°C on shaker at 180 rpm for 5 days so as to achieve complete degradation of feather by the keratinase produced, and the product was dried to produce KTF. Six soil amendments made in study included T1 (soil sample (1 kg.)), T2: soil sample (900 g) + 100 g of farm yard manure (FYM), T3: soil sample (900 g) + 100 g of vermicompost (VC), T4: soil sample (900 g) + 100 g of KTF, T5: soil sample (800 g) + 200 g of KTF and T6: soil sample (700 g) +300 g. of KTF adopting method of Hadas and Portnoy 1994. Soil parameters analyzed included using methods are soil moisture retention 930.15 (AOAC 2000), electrical conductivity (Jackson 1973), soil organic matter (Nelson and Sommers 1982), $CO_2$ evolution (Anderson 1982), available nitrogen (988.05 of AOAC 2000), phosphorous, potassium and microelements (Clement et al. 1995) and microbial counts—bacterial, fungal and actinomycetes by spread plate method.

## 3 Results and Discussion

The KTF was produced by semisolid state fermentation where complete degradation of feather was achieved in 5 days. After completion of the period, KTF was pooled and subjected to autoclaving at 10 lb/$inch^2$ for 15 min. The product was then cooled to room temperature and dried at 40°C and powdered. This KTF was blended in size by passing through 200 micron mesh sieve. Subsequently, homogenized KTF was stored in airtight containers Fig. 1.

Soil amendments were made in order to test the effect of addition of KTF to soil in terms of improvement in soil texture and available nutrients. The KTF amendments in the range of 10–30% were compared with farm yard manure (FYM) and vermi-compost (VC) in addition to unamended soil control. The amended samples were

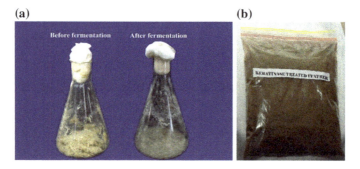

**Fig. 1** KTF powder preparation **a** SSF product **b** dried powder

maintained in triplicates and incubated at room temperature up to 75 days. Samples drawn at six periodic intervals between 0 and 75 days were analyzed to determine various soil parameters. Soil amended with 30% KTF showed higher moisture content of 59% up to 75 days which was similar to that of vermicompost amendment; whereas, other controls had lower water retention activity in the range of 50–52% (Fig. 2). A 1% increase of soil organic matter is estimated to increase approximately 3.7% water holding capacity of soil. Increase in water retention positively influences the crop yield of a soil by increasing nutrients availability for plant growth (Glaser et al. 2002). Soil EC gradually increased with incubation time and the magnitude of increase was higher in 20 and 30% KTF amended soils when compared to the controls followed by FYM and VC indicating good stabilization of soil (Table 1). All the KTF amendments showed increased nitrogen content from 0th day to 45th day after which it showed a decreasing trend. T6 amendment recorded maximum nitrogen content of 126 kg/ac. followed by T5 (106 kg/ac.), T4 (92 kg/ac.), T3 (88 kg/ac.) and T2 (86 kg/ac.), respectively.

All the KTF and organic amendments showed a rapid increase in phosphorous content upto15th day which continued gradually up to 45 days after which it showed a decreasing trend. Thirty percent KTF amendment recorded maximum phosphorous content (98 kilo/acre) followed by 20% KTF (96 kilo/acre) and 10% KTF (93 kilo/acre). FYM and VC showed 83 kilo/acre and 88 kilo/acre by 45th day. Soil

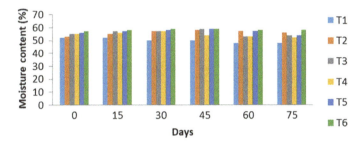

**Fig. 2** Comparison of moisture retention capacity in soil amendments

**Table 1** Change in EC among various soil amendments

| Amendment | Soil incubation (Days) | | | | | |
|---|---|---|---|---|---|---|
| | 0 | 15 | 30 | 45 | 60 | 75 |
| | *Soil electrical conductivity (MS/cm)* | | | | | |
| T1 (Control) | 0.64 | 0.62 | 0.14 | 0.18 | 0.13 | 0.05 |
| T2 (FYM) | 0.68 | 0.18 | 0.15 | 0.12 | 0.12 | 0.05 |
| T3 (VC) | 0.66 | 0.13 | 0.18 | 0.12 | 0.11 | 0.09 |
| T4-10% KTF | 0.68 | 0.11 | 0.26 | 0.28 | 0.23 | 0.04 |
| T5-20% KTF | 0.71 | 0.12 | 0.28 | 0.22 | 0.25 | 0.05 |
| T6-30% KTF | 0.71 | 0.10 | 0.28 | 0.30 | 0.30 | 0.08 |

amendment with organic manures shows gradual increase of potassium content up to 45th day after which it shows a gradual decrease up to 75 days.

KTF amendments showed increase of zinc content from 0th hour to 45 days then showed decreased trend up to 75 days. Thirty percent KTF amended soil was observed to have the highest amount of zinc content in soil. Soil amended with organic manures showed 6.2 ppm and 6.48 ppm of manganese contents. KTF amended soil showed much higher amount of manganese content. All the organic amendments including KTF amendments showed increase of manganese content up to 45th day and later showed decline. Highest level of iron content was noticed in T6 followed by T5 and T4, respectively, where maximum of 46.80, 41.26 and 36.04 ppm were observed (Fig. 3).

Organic matter content is greater especially in the topsoil where most of the bioactivity takes place (Choi and Nelson 1996). Soil organic matter slowly increased in various organic amendments during the experimental period upto 45 days (Fig. 4). On an average, T5, T6 and FYM were found to have higher soil organic matter followed by T4 and T3. Control showed least soil organic matter content. Organic matter content in the soil increased up to 45th day, after which there was a slight decrease.

Bacteria count was observed to increase significantly on the addition of amendments as compared to control up to 60 days (Table 2). The 0th day count of bacteria in CFU/g for T1 was $5 \times 10^7$ for control, and for FYM and VC it was in the range of $5 \times 10^8$–$6 \times 10^8$, respectively. KTF amendments showed significant increase

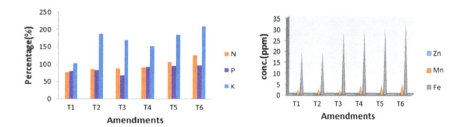

**Fig. 3** Available NPK and micronutrient contents in various soil amendments

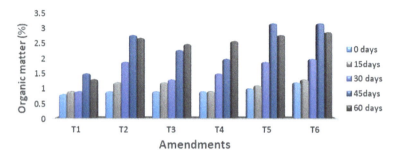

**Fig. 4** Organic matter content of various amendments

**Table 2** Viable counts with various Amendments

Soil incubation (Days)

| Amendment | Bacteria | | | Fungi | | | Actinomycetes | | |
|---|---|---|---|---|---|---|---|---|---|
| | 0 | 30 | 60 | 0 | 30 | 60 | 0 | 30 | 60 |
| | CFU/g of soil | | | | | | | | |
| T1 | $5 \times 10^7$ | $9 \times 10^7$ | $8 \times 10^7$ | $8 \times 10^3$ | $30 \times 10^3$ | $26 \times 10^3$ | $1 \times 10^2$ | $3 \times 10^3$ | $5 \times 10^3$ |
| T2 | $6 \times 10^8$ | $12 \times 10^8$ | $11 \times 10^8$ | $9 \times 10^3$ | $31 \times 10^3$ | $33 \times 10^3$ | $2 \times 10^2$ | $4 \times 10^3$ | $5 \times 10^3$ |
| T3 | $5 \times 10^8$ | $12 \times 10^8$ | $15 \times 10^8$ | $6 \times 10^3$ | $24 \times 10^3$ | $20 \times 10^3$ | $1 \times 10^2$ | $7 \times 10^3$ | $6 \times 10^3$ |
| T4 | $9 \times 10^8$ | $16 \times 10^8$ | $19 \times 10^8$ | $14 \times 10^3$ | $39 \times 10^3$ | $35 \times 10^3$ | $3 \times 10^2$ | $9 \times 10^3$ | $7 \times 10^3$ |
| T5 | $11 \times 10^8$ | $19 \times 10^8$ | $20 \times 10^8$ | $20 \times 10^3$ | $42 \times 10^3$ | $41 \times 10^3$ | $5 \times 10^2$ | $12 \times 10^3$ | $9 \times 10^3$ |
| T6 | $9 \times 10^8$ | $14 \times 10^8$ | $19 \times 10^8$ | $16 \times 10^3$ | $41 \times 10^3$ | $39 \times 10^3$ | $4 \times 10^2$ | $11 \times 10^3$ | $10 \times 10^3$ |

in bacterial count up to45th day after which it showed slight decrease up to 75th day. The counts were still significantly higher than control and FYM. Thus, KTF amendments were found to support significantly higher bacterial growth indicating sustained release of nutrients. The fungal populations enumerated in different amendments are given in Table 2. The counts also showed gradual increasing trend from the day of amendment till 45th day, after which there was a marginal decrease up to 75th day. KTF amendments showed similar pattern of higher fungal population up to 45th day as compared to other organic amendments and control indicating good support for fungal growth. Actinomycetes number also increased up to 45th day in control as well as in amendments after which a gradual decrease was observed up to 75th day (Table 2). Twenty and 30% KTF amendment showed maximum number of actinomycetes count followed by FYM and VC. The microbial count which is an indicator of microbial metabolic activity showed an increasing trend with KTF amendment.

Keratinase treated feather (KTF) had been reported to have better nutritive value in terms of total amino acid content and concentrations of cysteine, serine and methionine (Suneetha and Lakshmi 2004; Jeevana Lakshmi 2008). There was a considerable increase in the availability of free nitrogen (3.82–4.02 g/kg) in the keratinase treated feather, as compared to heat/acid treated (0.15 g/kg) or trypsin digested feather (1.5 g/kg). (Jeevana Lakshmi and Lakshmi 2015a). Both the digestibility and amino acid balance of feather meal were found to be improved by keratinases. Comparison of in vitro digestibility also showed that keratinase treatment resulted in ~ 2–2.5 fold increase in digestibility as compared to commercial feather meal. A significant ~2.5 fold increase in the percentage of proline and glycine content, a ~ 2 fold increase in cystine and ~1 fold increase of lysine and methionine were observed in KTF as compared to ones produced by other traditional treatments (Jeevana Lakshmi and Lakshmi 2015b).

# 4   Conclusion

Some of the commonly used organic fertilizers are bovine dung and urine, sheep manure, poultry waste, chicken manure, night soil, composted agricultural wastes, bat guano, vermicompost, etc. Among these, FYM, VC and bat guano are widely used as organic manure to improve soil fertility (Edwards and Arancon 2004; Lenin et al. 2010). However, it is still observed that the vegetables and fruits grown using organic means are sold in market at almost double the price compared to those produced by using chemical fertilizers making them unaffordable to common public (Nagavellamma et al. 2004).

Active organic matter provides habitat and nutrients for beneficial soil organisms that in turn help in building soil structure and its health. Organic amendments are emerging as an environmentally friendly alternative to the use of chemical fertilizer. KTF amendment of soil was observed to significantly improve the soil parameters, nutrient availability and had significantly higher soil organic matter (SOM) followed by other organic amendments. Thus, the present study highlights the application potential of this indigenously developed nutrient-rich keratinase treated feather which is an odorless, free running, eco-friendly, low-cost organic fertilizer leading to generation of wealth from poultry waste.

# References

Allpress, J. D., Mountain, G., & Gowland, P. C. (2002). Production, purification and characterization of an extracellular keratinase from *Lysobacter* NCIMB 9497. *Letters of Applied Microbiology, 34,* 337–342.

Anderson, J. P. E. (1982). Soil respiration. In: A. L. Page (Eds.), *Methods of soil analysis. Part 2. Chemical and microbiological properties. Agronomical Monograph* (pp. 831–866). Madison: ASA.

AOAC. (2000). *Official methods of analysis of AOAC international* (17th ed., pp. 240–296). USA: Gaithersburg.

Bach, E., Lopes, F. C., & Brandelli, A. (2015). Biodegradation of α and β-keratins by gram-negative bacteria. *International Biodeterioration and Biodegradation, 104,* 136–141.

Bockle, B., Galunsky, B., & Mueller, R. (1995). Characterization of a keratinolytic serine proteinase from *Streptomyces pactum* DSM 40530. *Applied and Environmental Microbiology, 61*(10), 3705–3710.

Brandelli, A., Sala, L., & Kalil, S. J. (2015). Microbial enzymes for bioconversion of poultry waste into added-value products. *Food Research International, 73,* 3–12.

Brandelli, A., Daroit, D. J., & Riffel, A. (2010). Biochemical features of microbial keratinases and their production and applications. *Applied Microbiology and Biotechnology, 85*(6), 1735–1750.

Choi, J. M., & Nelson, P. V. (1996). Developing a slow-release using poultry feathers. *Journal of the American Society for Horticultural Science, 121*(4), 634–638.

Edwards, C. A., & Arancon, N. Q. (2004). 18 the use of earthworms in the breakdown of organic wastes to produce vermicomposts and animal feed protein. *Earthworm Ecology, 2,* 345–355.

Farag, A. M., & Hassan, M. A. (2004). Purification, characterization and Immobilization of a keratinase from *Aspergillus oryzae. Enzyme and Microbial Technology, 34,* 85–93.

Friedrich, A. B., & Antranikian, G. (1996). Keratin degradation by *Fervidobacterium pennivorans,* a novel thermophilic anaerobic species of the order Thermotogales. *Applied and Environmental Microbiology, 62,* 2875–2882.

Glaser, B., Lehmann, J., & Zech, W. (2002). Ameliorating physical and chemical properties of highly weathered soils in the tropics with charcoal—A review. *Biology and Fertility of Soils, 35*(4), 219–230.

Gradisar, H., Kern, S., & Friedrich, J. (2000). Keratinase of *Doratomyces microsporus. Journal of Applied Microbiology and Biotechnology, 53,* 196–200.

Grazziotin, A., Pimental, F. A., de Jong, E. V., & Brandelli, A. (2006). Nutritional improvement of feather protein by treatment with microbial keratinase. *Animal Feed Sciences and Technology, 126,* 135–144.

Gupta, R., & Ramnani, P. (2006). Microbial keratinases and their prospective applications: An overview. *Applied Microbiology and Biotechnology, 70*(1), 21–33.

Hadas, A., & Portnoy, R. (1994). Nitrogen and carbon mineralization rates of composted manures incubated in soil. *Journal of Environmental Quality, 23*(6), 1184–1189.

Jackson, M. L. (1973). *Soil chemical analysis* (2nd ed., pp. 59–67). New Delhi, India: Prentice Hall of India Private Limited.

Jayathilakan, K., Sultana, K., Radhakrishna, K., & Bawa, A. S. (2012). Utilization of byproducts and waste materials from meat, poultry and fish processing industries: a review. *Journal of Food Science and Technology, 49*(3), 278–293.

Jeevana Lakshmi, P., & Lakshmi, V. V. (2015a). Enhancement in nutritive value and invitro digestability of keratinse treated feather meal. *Journal of Scientific & Engineering Research, 6*(2), 36–40.

Jeevana Lakshmi, P., & Lakshmi, V. V. (2015b). Evaluation of degradative products of feather degradation by *Bacillus* sp. *International Journal of Scientific & Engineering Research, 6*(2), 330–333.

Jeevana Lakshmi, P., Kumari, Ch. M. and Lakshmi, V. V. (2013). Efficient degradation of feather by Keratinase producing *Bacillus* sp. *International Journal of Microbiology* (p. 7). http://dx.doi.org/10.1155/2013/608321. Article ID 608321.

Jeevana Lakshmi, P. (2008). *Fermentative production of keratinase by Bacillus sp. and its relevance to recycling of poultry feather waste* (Ph.D. thesis). Submitted to Sri Padmavati Mahila Visvavidyalayam, Tirupati.

Kumari, Ch M, Jeevana lakshmi, P., & VV, Lakshmi. (2015). Microbial keratinases and their applications. *International Journal of Scientific & Engineering Research, 6*(2), 50–54.

Kumari, Ch M, & Lakshmi, V. V. (2015). Fermentative production of keratinase using solid agricultural wastes. *International Journal of Scientific & Engineering Research, 6*(2), 56–57.

Kumari, Ch. M. (2011). *Production of microbial keratinases and its application in bioremediation of feather* (Ph.D. thesis). Submitted to Sri Padmavathi Mahila Visvavidyalayam, Tirupati.

Lenin, M., Selvakumar, G., & Thangadurai, R. (2010). Growth and nutrient content variation of groundnut *Arachis hypogaea* L. under vermicompost application. *Journal of Experimental Sciences, 1*(8), 210–215.

Lin, X., Lee, C. G., Casale, E. S., & Shih, J. C. (1992). Purification and characterization of a keratinase from a feather-degrading *Bacillus licheniformis* strain. *Applied and Environmental Microbiology, 58*(10), 3271–3275.

Govern, Mc. (2000). Recycling poultry feathers: More bang for the cluck. *Environmental Health Prospective, 108*(8), 366–369.

Moritz, J. S., & Latshaw, J. D. (2001). Indicators of nutritional value of hydrolyzed feather meal. *Poultry Science, 80,* 79–86.

Nagavellamma, K. P., Wani, S. P., Stephane, L., Padmaja, V. V., Vineela, C., Babu Rao, M., Sahrawat, K. L. (2004). *Vermicomposting: recycling wastes into valuable organic fertilizer. Global theme on agroecosystems* (Report No. 8), 20–28.

Nelson, D. W. and Sommers, L. (1982). Total carbon, organic carbon, and organic matter. *Methods of Soil Analysis. Part 2. Chemical and Microbiological Properties, (methodsofsoilan2)*, 539–579.

Onifade, A., Al-Sane, N., Al-Musallam, A., & Al-Zarban, S. (1998). Potentials for biotechnological applications of keratin degrading microorganisms and their enzymes for nutritional improvement and others keratins as livestock feed resources. *Bioresource Technology, 66,* 1–11.

Papadopoulos, M. C. (1984). *Feather meal: Evaluation of the effect of processing conditions by chemical and chick assays*. (Ph.D. thesis). Agricultural University, Wageningen, Netherlands.

Clement, R. E., Eiceman, G. E., & Koester, C. J. (1995). Environmental analysis. *Analytical Chemistry, 67*(12), 221–255.

Santos, R. M. D. B., Firmino, A. A., de Sai, C. M., & Felix, C. R. (1996). Keratinolytic activity of *Aspergillus fumigatus* fresenius. *Current Microbiology, 33,* 364–370.

Suh, H. J., & Lee, H. K. (2001). Characterization of a keratinolytic serine protease from *Bacillus subtilis* KS-1. *Journal of Protein Chemistry, 20,* 165–169.

Suneetha, V., & Lakshmi, V. V. (2004). Optimization of fermentation parameters for hair degrading microorganisms isolated from Tirumala hills. *Asian Journal of Microbiology, Biotechnology & Environmental Sciences, 6,* 231–233.

# Conversion of Waste Polythene Bags/Wrappers into Useful Products

Priskila Macwan

**Abstract** Solid waste management and conversion of waste polythene bags/wrappers is the proposed pilot by World Vision India jointly with Green for Life Foundation, as the unprecedented use of polythene bags has a huge negative impact on the ecosystem. We have developed the concept of social enterprise through recycling of polythene bags under Government of India's "Swatchh Bharat Mission" and using the technology of Green for Life Foundation, in Jain Kunj slum (ward no. 80) which is fragile in context of adaptation to climate change and socioeconomically most backward of the lot. The problem of waste management is rampant and cross-cutting across the entire wards. The proposed pilot is design to achieve three goals under Sustainable Development Goals, they are

Goal 1—No poverty
Goal 8—Decent work and economic growth, and
Goal 13—Climate change.

Our pilot is a uniquely designed program to fit the needs of our community.

- The purpose of conversion of waste polythene bags/wrappers is to keep these polythene bags/wrappers off the ground and prevent them from getting into the soil.
- To collect and convert waste polythene bags/wrappers that will minimize environmental hazards.
- To convert waste polythene bags/wrappers through the use of low-cost alternative technology.
- To conduct training courses and workshops for rag pickers, rag segregators, and manufactures both on soft skill and technical skills.

Develop Social Enterprise—The pilot helps empower the underprivileged and marginalized people in our community by creating, influencing environmental and socioeconomic development programs thus resulting in—Building a Resilient Community. Our goal is to promote and encourage establishing community-based Social Enterprise Unit focused on converting waste polythene bags into useful products

P. Macwan (✉)
World Vision India, Kolkata, India
e-mail: priskila_macwan@wvi.org

© Springer Nature Singapore Pte Ltd. 2020     77
S. K. Ghosh (ed.), *Waste Management as Economic Industry Towards Circular Economy*,
https://doi.org/10.1007/978-981-15-1620-7_5_8

through hands-on and minds-on activities, teamwork and community involvement, for creating a sustainable Model Slum.

**Keywords** Solid waste management · Conversion · Technology · Polythene bags · Swatchh Bharat Mission · Social enterprise · Resilient community · Single use fusion process · Community ownership · Women empowerment · Community economy

## 1 Introduction

World Vision is one of the world's leading child-focused humanitarian organizations. Through development, relief and advocacy, we pursue fullness of life for every child by serving the poor and oppressed regardless of religion, race, ethnicity, or gender as a demonstration of God's unconditional love for all people.

In order to enable families to enhance income and thereby livelihood, World Vision India promotes interventions in the areas of agriculture (production, value addition, food processing, organic farming, irrigation, water development, product marketing, etc.), livestock development, enhancement of market access, skill upgradation of unemployed youths, and community empowerment.

Solid waste management through conversion of polybags/wrappers is uniquely designed pilot program to fit community World Vision India work with in Kolkata, by empowering the underprivileged and marginalized people of society having an important role to play in creating, influencing, and participating in environmental and socioeconomic development programs. By taking this initiative forward, non-formal skills training for the urban community to help them generate economic so that these communities become capable of "bouncing back" from adverse situations by actively influencing and preparing for economic, social, and environmental change— Building Resilient Community making Model Slum.

## 2 About Technology: Energy & Emission

The carbon footprint of plastic (LDPE or PET, polyethylene) is about 6 kg $CO_2$ per kg of plastic. Polyethylene PE is the most commonly used plastic for plastic bags.

The process of conversion of waste polybags/wrappers is a single-stage fusion process, and the prescribed equipment uses standard 240 V. The energy used during this process is the lowest among all types of plastic conversion, though waste polybags/wrappers are considered "unrecyclable" in the current plastic recycling industry which uses standard processes*.

The energy used during the process adopted by us is about 0.90 KWH or 3087 heat units per hour for recycling 600 g of waste polybags. To process 1 kg of waste polybags, energy equivalent to 1.5 KWH or 5145 heat units per hour will be used.

The objective is

- The purpose of conversion of waste polythene bag is to keep them off the ground and prevent them from getting into the soil by collecting and converting waste polythene bag into useful products that will minimize environmental hazards
- To convert waste polythene bags (HDPE, LDPE & PP) through the use of low-cost alternative technology called fusion technology
- Develop social enterprise—To create an income generating opportunity for poor people through conversion of waste polythene bags into useful products.

## 3  Literature Survey

As per World Economic Forum Reports, by the year 2050, there will be more plastic waste, by weight, in our oceans than there are fish. On April 4, 2013, The Supreme Court of India said "We are sitting on a plastic bomb," in response given by Central Pollution Control Board. As per Central Pollution Control Board, Kolkata generates about 425.7 tonnes of plastic waste per day. As 40% of plastic waste is not recycled, the daily collection to untreated plastic waste is 170 tonnes. Though the Kolkata Municipal Corporation (KMC) introduced its first scientific solid waste compactor station and modern portable compactors in various locations, generally its use is for landfilling and on top of that still there is huge tone of plastic remaining to collect while waste collection. Thus, it needs immediate support and implementation.

The urban poor is vulnerable to climate change often living in riskier urban environments such as floodplains or unstable slopes, working in the informal economy, and with fewer assets—are most at risk from exposure to hazards. And it has analytically identified the 9 most vulnerable wards that may need specific attention in designing adaptation strategies in Kolkata. Among them, Ward 80 falls under vulnerable category.

Jain Kunj slum (ward no. 80) is fragile in context of adaptation to climate change and socioeconomically most backward of the lot. The problem of waste management is rampant and cross-cutting across the entire ward, and Jain Kunj is on port area surrounded by Hooghly River.

Solid waste management and conversion of waste polythene bags/wrappers into useful products is the proposed pilot by World Vision India jointly with Green for Life Foundation, where World Vision India developed the concept of social enterprise through recycling of polythene bags under Government of India's "Swatchh Bharat Mission" and using the technology of Green for Life Foundation.

Our Goal is to promote and encourage establishing community-based Social Enterprise Unit focused on converting waste polythene bags into useful products through hands-on and minds-on activities, teamwork and community involvement, for creating a sustainable Model Slum—resulting in Building Resilient Community for a healthier and greener society, which are key to poverty reduction.

# 4   The Work

This project involves various aspects of conversion of waste polybags/wrappers. A detailed description of the various stages involved in the conversion process is mentioned below:

A. **Promotion and awareness**:

The promotion and awareness of waste polybags is much required as there is a huge gap lying in the knowledge of people how to handle polywaste. At present, Jain Kunj community is filth with polywaste which is easily identified by observing their land and gutter lanes. During heavy rain, Jain kunj experiences waterlogging for days which disturbs the regular life of people residing there. Before we take any further step, we planned 3 days interactive awareness sessions for the community, where in each session we reached out to max 50 people at a time with hope that these 150 people to reach out to other community people with same massage.

Also, having such interactive sessions are for the future support where community understating how the whole process will undergo recycling polywaste and empowered community-level economic development. We also observed and identified interested 15 women for our further intensive training who will be part of workshop as experts to recycle.

B. **Formation of livelihood-based collective**:

Small businesses or small enterprises are key to poverty reduction because they provide an opportunity for people who are poor to earn the income required to purchase goods and services to ensure well-being and survive shocks. Apart from this category, we have workers working in different segments of informal economy who also need to focus to avail benefits and services. Through this enterprise, we aim to build the capacity and confidence of the 15 community members themselves to create an environment where it is easier for people who are poor to increase and stabilize their income. It facilitates collective action by micro-enterprise owners to improve their businesses, create jobs, and open up growth opportunities. This enterprise will increase their income, assets, and confidence of poorer families, through developing their skills and knowledge, building relationships with other business stakeholders in their communities, and enabling a diversity of enterprises to develop in the target area in below given areas,

Forming Producer Collective

- Step 1: Form identified members as producers' collective
- Step 2: Sensitize producers on benefits of producers' organization
- Step 3: Select the office bearers and orient them on their roles and responsibilities
- Step 4: Develop by-laws of producers' organization
- Step 5: Register producers' company
- Step 6: Enroll members in producers' company.

Train members and office bearers on operational, sustainable management of the producer collective

- Step 1: Train on primary members and their role, second level at office bearers both operational and financial sustainability
- Step 2: Train on network with banks for their cooperation which provided hassle-free loans to the PC for working capital
- Step 3: Train them on value chain on their production business opportunity that gives high values.

Train on specific business services to its members (on fee basis to have maintain operational sustainability)

- Saving services
- Input services—Seeds/collectively using common assets of high value
- Market-related services—Production, harvesting, processing, procurement, grading, pooling, handling, marketing, selling, and export of primary produce of the members or import of goods or services for their benefit:
- Financial services—linkages to bank and government schemes and pass on the benefit
- Welfare services to its members.

C. **Collection of waste polybags/wrappers**:

Collection will be done from within community to ensure Jain Kunj is polywaste-free community and from other identified location like offices, malls, factories from where it is easy to find bulk polywaste. Within the community, we will place several bins and for other locations, we will arrange pickup for fortnight. Follow-up on collection will be carried out by the collection team.

D. **Segregation of waste polybags/wrappers**:

Once the waste polybags/wrappers are collected, they will be segregated into categories such as High-density polythene (HDPE), low-density polythene (LDPE) and polypropylene (PP), small and big waste polybags/wrappers. They are to cut open, wash in detergent water, disinfected, and then dried.

E. **Conversion of waste polybags**:

The dried waste polybags/wrappers are placed in the conversion machine by applying regulated heat and temperature. These waste polybags are processed and converted into sheets of various size and thickness. Thereafter, these sheets can be designed into useful products such as folders, files, document/pencil pouches, table mats, and TV/computer covers.

F. **Marketing**

Marketing and promotion is an integral feature for creating a sustainable and viable project. By adopting an effective, systematic marketing strategy, we propose to market these recycled products through direct and retail marketing.

## 5   Target Customers

a. General public
b. Corporate
c. Shops/stores
d. Online stores
e. Schools
f. Institutions.

## 6   Approach

Project will provide close on-site support, skill training and self-employment support, handholding assistance for financial literacy, enterprise management, and other developmental support and facilitating linkages to corporates, industries, and organizations for ongoing marketing. Intensive training, skill development, technical support, and aggressive marketing are the keys to make this project a viable income generating and community development project.

## 7   Methodology

- Local ownership—To ensure local ownership through collectives or groups and problem framing at the grassroots level.
- Empowerment—To empower the individual and groups in order to help own self, take decisions, and make changes.
- Sustainability—To ensure the long-term social, economic, and environmental sustainability.

The processed waste poly bags are then converted into sheets in various sizes and thickness as per requirement. Thereafter, these sheets during the initial phase will be designed into useful products such as folders, files, document/pencil pouches, table mats, and TV/computer covers for which we have available market. In later stage, based on market, we will keep adding more designs with upgrading skills. Basically, conversion of polybag will give us raw material which we will use to produce various finished goods as well as the raw materials itself can be marketed to various vendors.

# 8    Role of Community

- Unit will be settled in community and will be under observation of local community-based organizations (CBO)
- Unit will be run by local selected, trained community people
- Related small groups under value chain also will be fulfilled by community people
- In later stage, unit will be legally registered under cooperative/trust registration.

# 9    Expected Results

Our goal is to promote and encourage establishing community-based Social Enterprise Unit through hands-on and minds-on activities, teamwork and community involvement, for creating a sustainable Model Slum for a healthier and greener society, which are key to poverty reduction. This pilot is to increase the income, assets, and confidence of poorer families, through developing their skills and knowledge, building relationships with other business stakeholders in their communities, and enabling a diversity of enterprises to develop in the target area (The New Plastics Economy 2016).

# 10    Conclusions or Recommendations

Waste management is need of an hour as it is an alarming issue not of one country but for the whole world. Such low-cost solutions not only help in managing waste but also create opportunity to create livelihood for many poor families. Waste management technology and community development can go hand in hand and thus both the parties need to come together and create many such models.

# 11    Future Scope of Work

We will link with corporates, industries, and organizations for collection of plastic and marketing eco-friendly recycling product out of polybag. Linkages with Kolkata Municipal Corporation (KMC) to be part of such intervention and adopting as solution to waste management.

The similar program we would also like to scale up in other six cities where World Vision India's program are called as My city Initiatives in Bangalore, Chennai, Delhi, Guwahati, Hyderabad, and Mumbai to address polywaste and urban vulnerable poor which can consider in upcoming such funding opportunity.

**Acknowledgements** The author appreciates Green for Life Foundation's objective of identifying and promoting activities which improve and protect the ecology of Planet Earth and fragile environment, and it is committed to think and act green, came up with such low-cost technology which is easy to adopt by slum community, and wishes them to come up with many more solutions to waste management.

# Reference

*The New Plastics Economy: Rethinking the future of plastics.* Report Published on 19th January 2016. https://www.weforum.org/reports/the-new-plastics-economy-rethinking-the-future-of-plastics.

# Upcycling of Scraps from Technical Institutes: A Case Study—Govt. Industrial Training Institute Berhampur

**Rajat Kumar Panigrahy**

**Abstract**  Govt. ITI Berhampur having 3600 students is one of the largest ITIs in the country. The trainees of the ITI developed this innovative method of Upcycling of scrap by managing waste beyond skill development. An integration of skilling under Skill India Mission and Swachh Bharat to make the environment clean, green and safe from the iron/aluminum/electronic/automobile scraps that are released and left out during their regular training practice creating the environment hazardous is the most desired need of the hour. Normally, in a technical training institute, the trainees go for filing/drilling turning chipping practices which generate various kinds of iron scraps. Also, electrician/electronic trainees produce scrap from PVC wire/electronic component and waste PCBs. Automobile trainees leave the discarded spare parts from two wheelers and four wheeler scrap. Around 1000 kg of scrap is produced during the training session at Govt. ITI Berhampur in every semester. Some iron-dust mixes with the soil and is washed off in the rainwater creating water pollution and soil pollution as well. Transporting these items to the dumping grounds or the landfill areas is also not a solution as it is neither economical nor eco-friendly. However, by this innovative approach "upcycling of Scrap", 50–60% of items can be recycled. These recycled finished products not only are the answer to the biggest challenge of scrap disposal, but they also have immense market value. Around 13,105 ITIs (both government and private) in the country generate a gigantic amount of scrap every day. To recycle these scrap items & to convert them into scrap designs, thus, giving beauty and value to the undesired products which can be placed as objects adding beauty is what the "Recycling of Scrap" aims toward. Going by the facts, it has been identified that out of the total materials used as a part of the practical training curriculum, 30% of the total materials used turns out as waste material post-production of jobs. The ITI trainees produce some beautiful "sculpture" from these scraps. By this, they enhance their learning skills like welding, painting, filling turning, welding and fitting. This helps them to enhance their skill as well as the "sculpture" developed is immense market value. It is sold like hot cake to decorate at houses by the interior decorators. This is a new era of skill development. By this, the trainees earn when they learn and helps to clean the environment.

R. K. Panigrahy (✉)
Industrial Training Institute, Berhampur, India
e-mail: muna2867@gmail.com

© Springer Nature Singapore Pte Ltd. 2020
S. K. Ghosh (ed.), *Waste Management as Economic Industry Towards Circular Economy*,
https://10.1007/978-981-15-1620-7_5_9

**Keyword** Scraps · Upcycle Recycles sculptures

# 1   Introduction

Considering the fact that there are 13,105 ITIs in India as per 2016 data consisting of both government and private ITIs, the total waste is far beyond expectations.

# 2   Literature Survey

Now, this was targeted as an entirely new dimension for a wider range of opportunity. As a part of constant improvement and development, we at Govt. ITI Berhampur have identified and implemented the "Technical Institute Waste Management" system. There are approximately 2500 students at ITI Berhampur enrolled under multiple trades. As a part of the education system, at ITI Berhampur, a lot of practical work is done by the students to be skilled workers which is the demand of the nation now. Considering all these facts, ITI Berhampur has taken the initiative in creating a change in the whole process of waste management (Tables 1, 2, 3 and Figs. 1, 2, 3, 4).

We have developed a skillful use of these scrap materials to give a new definition to the so-called scrap. Above all, our process of "waste management", within the technical institute itself, is creating a new era of skill development by using the

**Table 1**  Last four year mechanical trade trainees of ITI, Berhampur

| Sl. No. | Trade name | 2015 | 2016 | 2017 | 2018 |
|---------|------------|------|------|------|------|
| 1 | Fitter | 252 | 756 | 756 | 1008 |
| 2 | Turner | 48 | 96 | 96 | 96 |
| 3 | Machinist | 48 | 96 | 96 | 96 |
| 4 | Foundry Man | 21 | 21 | 21 | 21 |
| 5 | Welder | 48 | 126 | 126 | 126 |
|   | Total | 417 | 1095 | 1095 | 1347 |

**Table 2**  Last four year automobiles trade trainees of ITI, Berhampur

| Sl. No. | Trade name | 2015 | 2016 | 2017 | 2018 |
|---------|------------|------|------|------|------|
| 1 | COE, automobile | 252 | 252 | 252 | 252 |
| 2 | Mech. M.V. | 21 | 63 | 63 | 63 |
| 3 | Mechanic tractor | 21 | 21 | 21 | 21 |
| 4 | Mechanic diesel |  | 63 | 63 | 63 |
|   |  | 294 | 393 | 393 | 393 |

**Table 3** Last four year electrical and electronics trade trainees of ITI, Berhampur

| Sl. No. | Trade name | 2015 | 2016 | 2017 | 2018 |
|---|---|---|---|---|---|
| 1 | Electrician | 252 | 756 | 756 | 756 |
| 2 | Electronics mechanic | 63 | 126 | 126 | 126 |
| 3 | Instrument mechanic | | 21 | 42 | 42 |
| | | 315 | 903 | 903 | 903 |

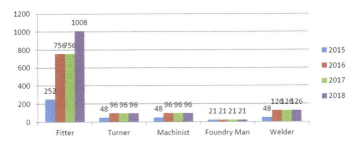

**Fig. 1** Last four years mechanical trade trainees' data

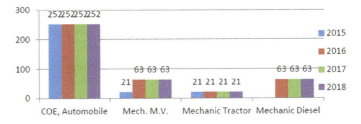

**Fig. 2** Last four years automobiles trade trainees' data

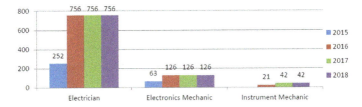

**Fig. 3** Last four year electrical and electronics trade trainees of ITI, Berhampur

waste materials. The fact that around 1000 kgs of scrap is disposed to the environment every month by a single technical institute, the total amount of scrap being disposed every day throughout the country by these technical institutes is alarming and the environment disposal of this industrial waste is the biggest challenge.

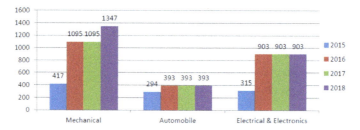

**Fig. 4** Last four year trainees of mechanical, automobile and electrical sector trainees

## 3 The Work

This positive change had to face many challenges before turning out to be positive outcome. There was this challenge of handling such huge volume of waste that needed a separate and dedicated team which was quite impossible with the existing manpower. Also, the waste could not be disposed of anywhere because of the probable threats to health and environmental pollution. There was minute sharp and tiny waste that could cause injury. These materials also when getting mixed up with soil caused the loss of fertility. Due to this, the vegetation cover will be getting affected badly which is the major concern in today's modern world. Also in the long run, this will also become a reason for global warming. Initially, there seemed to be never ending challenges in putting up this herculean task into action as the coordination of different trades during the layout of the process was also a big question. Then identifying the appropriate utilization of the scrap in waste transformation had its own difficulties, finding innovative use of the waste products and also coming together of best skills from multiple trades like turning, milling, filing, welding, brazing, painting and other trades.

**The Basic Challenges Can Be Summarized As**:

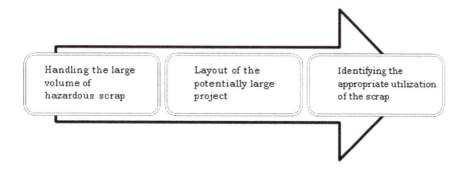

# 4 Approach and Methodology

We at the institute have added innovation to this scrap which has finally given birth to a new era of exposure called the—"Nasta Ru Shrestha" project. We have a number of trades within the institute and have, thus, adapted the sharing of applied knowledge concept from almost all the trades. We identified the effects of contribution of each trade in the final creation. The fitters, welders, electricians, plumbers, automobile, trainees all came under one roof of understanding and were working toward the concept of—"Nasta Ru Shrestha". Trainees from all the trades came together with the basic technical skills and created different models sculptures, which has given an entirely lively look to the so-called pieces which were initially considered as scrap.

For all this to be put up in a uni-direction positive output, first of all, an integrated team was created with the best hands from various trades like fitter, turner, machinist, welder, painter, plumber, electronics, MMV, electrician and every other trades also joined hands in this wonderful process. Focus was given on safety and hygiene measures at workplace. As a result of this continuous practice, the scattered disposal of waste was first limited, and thereby health and environmental hazard was avoided to a great extent. Next step taken was to identify different types of waste generated at workshops, and sorting of waste was the next step that was put into place. Maintenance of hygiene at workplace also minimized the risk from these scrap materials, and attention was provided to properly collect the waste from all the trades.

This strategy has a number of benefits to be precise. It has majorly contributed in solving the greatest challenge of Technical Institute Waste Management. This is an absolutely environment-friendly strategy and, thus, is totally eco-friendly. The most noticeable benefit of this concept is that it has opened the doors for a new area of skill development. The idea is to implement the non-conventional skills effectively giving birth to a new innovative creation.

We have a scrap museum at our ITI that showcases the innovation, efforts and skills of our students. This museum was inaugurated by Shri Subroto Bagchi—Chairman OSDA. The museum has hard-to-believe pieces of art made from junk. From home, décor items that will enhance the beauty of the house to equipment helpful in day-to-day work at construction places and other in-house equipment have been developed. Apart from waste management, these pieces of art which have not only added beauty to the junk, but the fact being these pieces of scrap have been converted into real usable finished products. The project which involved the ideas, concept, innovation and a whole-hearted dedicated work not only solved the problem of waste disposal, but has importantly given a hands-on entirely new area for skill development and utilization of skills before the deployment of these candidates. These ITI trainees, the major target now, had the platform to implement innovation, to the scrap. In doing so, they gathered a lot of work experience as it is majorly involved teamwork.

Thus, what can be rightly summarized is that the innovation and skills applied to the scrap had now resulted in a new usable finished product and a way of experienced learning before deployment. Talking of the benefits of this project, what is noteworthy here in the process is that it has contributed a lot to various government practices as well. A start-up toward an innovative solution to global warming is commendable. This project has a major contribution to the Swacch Bharat Abhiyan. In today's world, innovation is the key to achieve new milestones. But only having an innovative idea is worthless unless and until it is converted into action. At ITI Berhampur it was not like that, here innovation was brought into action. The best innovative use of materials that no one could have imagined off was penned down. Dedicated efforts were put forward in sketching the idea from multiple sources available like online referrals, industry leaders and other such sources. After preparing a sketch, the scrap materials were then identified that could possibly fit the purpose. Creativity saw new dimensions among the students, and there were unique ideas flowing in. Considering each and every unique idea, the best one was penned down. Finally, the idea was put into action with the coming together of teams from all the trades. After achieving the initial success in converting a scarp into an art, it was showcased as a unique art via an art gallery that was created especially for this purpose. The gallery was named as "Scrap Museum" and items those were once scrap now became a piece of art that attracted visitors from across regions and witnessed the skills and art of the students. This scrap museum was not just a simple display of items; it is an inspirational story that makes its visitors motivated on thinking that everything in this world has its use. Learners are now more enthusiastic and focussed because of this scrap museum. Everyone is keen on creating their own masterpiece, and in turn, they are putting in the best practical skills into action as a masterpiece also needs a master's touch. This practice of converting waste into best has now currently impacted the students across Berhampur ITI at first and also across various other ITIs at Odisha who have stepped into the path created by us. Perhaps this has really turned out to be the best way in which waste is managed, skill is enhanced during creating the art piece and most important part is that learners are now enjoying their work. And what best can happen when a learner is enjoying to work and put his learnings into action. In future, this project will definitely have a great positive impact on the learners as well as our society. If this project implemented across all ITIs in India, then definitely India is going to a better place.

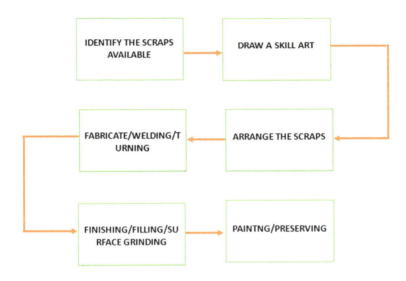

## 5 Methodology

This project can by large be considered as one which is a single point solution to a wide range of challenges:

- Waste management
- Innovative solution to global warming
- Applied conventional skills for producing better finished products
- New area of skill development
- Skill utilization and gaining work experience before deployment
- Wider dimensions for employment generation

The scrap materials and waste that were earlier creating a challenge in disposal and were hazardous are now such beautiful pieces of art that no one would ever want to dispose them off. Many of those visited the scrap museum so fascinated by these pieces of aft that they wanted to take away these items and use them to decorate their own house/offices/institutions, etc. This transformation of waste into best has also made the transformation of the students of the ITI. It feels psychologically satisfied in turning waste into best as it symbolizes that everything in this world holds its value, and the difference is how we look at it. This process has not only reduced the waste, but it has also made the students aware and alert about proper utilization of resources.

The students are also empowered to make their own "start-up" considering the use of these wastes, where they are exposed to a vast marketplace that holds a bright future as it is yet to be discovered. The Berhampur municipality which is in the race of Smart city now encourages the process. The district magistrate has inaugurated an exhibition for public awareness (Figs. 5, 6, 7 and 8).

**Fig. 5** Inauguration of exhibition on recycling of scrap under the theme Swachhata hi seva on 17 September 2018 by the district magistrate

**Fig. 6** Scripture from mechanical waste

**Fig. 7** Scriptures from automobile sector

**Fig. 8** Scriptures from E-waste

## 6  Significant Results

The ITI trainees produce some beautiful sculpture from these scraps by applying their own skill like welding, painting, filing and fitting. This helps them to enhance their skill as well as the sculptures have some market value. It is sold like hot cake to decorate at house by the interior decorators. This is a new era of skill development.

Comparison of the pre-deployment with post-deployment scenario, "Nasta Ru Shrestha", in itself is a concept which holds its essence in converting the waste to best. Creating new with new is not what the need of the hour is. Thus, we at Govt. ITI Berhampur believe in creating something new out of what is hidden but present in our environment. All most all the Govt. ITIs in the state are being directed to implement the process. This is achieved Skoch Order-of-Merit on 21 June 2018 under the training, skill development and innovation category. This can be extended to all the ITIs in the country (Fig. 9).

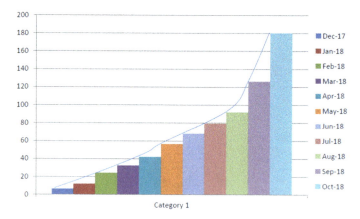

**Fig. 9**  Number of sculpture items prepared by the trainees of ITI, Berhampur

# 7  Conclusion

This is worth saying that our students are not just technical workers but they are artists who have created masterpieces by their skill and effort. We at ITI Berhampur are proud to have contributed in creating a pollution-free environment and also living by our Honorable Prime Minister's word of "Kaushal Bharat-Kushal Bharat".

# Reference

https://ncvtmis.gov.in/Pages/ITI/Search.aspx.

# A Business Model for Enterprise Resource Planning (ERP) by Establishing a Sustainable Supply Chain Management

Ipsita Saha, Amit Kundu and Sadhan Kumar Ghosh

**Abstract** *Purpose* The eco-friendly approach has become an active component of sustainable strategy for both academia and industry. The concept of green SCM is relevant as it integrates ecological wisdom of any organization with organizational supply chain actions. *Methodology* A pilot survey was done through framed questionnaire, and from value chain analysis, we have developed a model for sustainable strategy. The data was collected and analyzed to identify the major components of sustainability. The valid data was analyzed with the help of regression and correlation analysis. Findings: Basically, the rapid industrial expansion has led to greenhouse gas emissions; toxic pollutants those as a whole reflect declined sustainable growth. So, this study suggests that green HR management can be taken as new initiative by providing training toward employees on how to adopt sustainability practices from the view point of customer and suppliers. *Research implications* Sustainable supply chain process has theoretical implication and practical significance as well to ensure overall sustainable development for any industry. This model suggests that management support is always required for implementing the sustainability strategy in the organization. This novel business model guides the managers for implementing sustainable supply chain management practices in the organization.

**Keywords** Supply chain management · Sustainable supply chain process · Sustainable supply chain process

## 1 Introduction

In India, it becomes very difficult to sustain in the competitive environment. In the supply chain process, the eco-friendly strategy should be maintained. By 2050, the combined GDP of the emerging economies will be more than 50% of the total GDP of the world. This growth rate results the obvious emission of carbon footprint and

I. Saha (✉) · S. K. Ghosh
Department of Mechanical Engineering, Jadavpur University, Kolkata, India
e-mail: ipsita.saha49@gmail.com

A. Kundu
School of Management Studies, Techno India, Kolkata, India

© Springer Nature Singapore Pte Ltd. 2020
S. K. Ghosh (ed.), *Waste Management as Economic Industry Towards Circular Economy*,
https://doi.org/10.1007/978-981-15-1620-7_5_10

methane gas in the environment. Hollos et al. (2012) examined the implications of sustainable supplier cooperation and studied the influence of supplier cooperation on performance of the organization. The increasing importance of sustainable behavior in business has enhanced its impact on supply chain management. Firms foster sustainability in their supplier base in reaction to growing sustainability requirements in various ways, including sustainable supplier cooperation. Knowledge about the effects of sustainable supplier cooperation on firm performance is limited; therefore, this study tests antecedents and implications of sustainable supplier cooperation according to the triple bottom line (Gavronski et al. 2011). A survey of Western European firms reveals that sustainable supplier cooperation has generally positive effects on firm performance across social, green, and economic dimensions. However, only green practices have positive significant effects on economic performance, not social practices (e.g., child labor rules). Similarly, Gavronski, Klassen, Vachon, Nascimento, and Luis Felipe proposed that manufacturing's choice of environmental technologies is expected to be partly driven by the organizational context and receptivity to new ideas and innovation (Gavronski et al. 2011).

The present study makes three significant contributions. First, it explains the role of environment policy for enhancing the sustainability strategy on the supply chain management. Second, it refined and validated scales that capture organizational processes within operations which can enhance adoption of sustainability practices among suppliers. Finally, this research highlighted the important role that plant-level social climate has on fostering a greater emphasis on pollution prevention. The managerial implications of this research are twofold.

To establish a sustainable business model, organizations need to remodel the supply chain strategies with the advent of new technologies. Traditional supply chain management (SCM) means transferring goods and services. But modern aspect focuses on strategic SCM, where supply chains are used as a resource to satisfy increased customer demand and to enhance firm performance (Ketchen et al. 2007). IT practices and techniques are used to enable information sharing across supply chain partners, by integrating both internal and external business functions. In addition, the alignment of IT goals and objectives with strategic SCM can increase efficiency, productivity, and profitability. In this paper, we explore the impact of information technology (IT) on the development of competitive advantage throughout the supply chain.

## 2   Literature Review

If we consider Green SCM (GSCM) practices, Zhu and Sarkis (2004) divided GSCM into four major dimensions: internal environmental management, external environmental management, investment recovery, and eco-design. Internal environmental management includes commitment of GSCM by senior managers, support for GSCM by mid-level managers, cross-functional cooperation for environmental improvements, total quality environmental management, environmental compliance and auditing programs ISO 14001 certification, and environmental management systems. The external environmental management includes GSCM practices, providing design specification to suppliers that include environmental requirements for purchased item, cooperation with suppliers for environmental objectives, environmental audit for suppliers' internal management, suppliers' ISO14000 certification, second-tier supplier environmentally friendly practice evaluation, cooperation with customer for eco-design, cooperation with customers for cleaner production, and cooperation with customers for green packaging. Investment recovery includes investment recovery of excess inventories/materials, sale of scrap and used materials, and sale of excess capital equipment. Eco-design includes design of products for reduced consumption of energy, design of products for reuse, recycle, recovery of material, component parts, design of products to avoid or reduce the use of hazardous, and products and their manufacturing process. Vickery et al. (1999) defined five supply chain flexibilities based on previous operations literature in order to look at supply chain uncertainty problems. The author described the flexibility type including product flexibility, volume flexibility, new product flexibility, distribution flexibility, and responsiveness flexibility (Acar et al. 2010). Product flexibility is defined as the ability to customize product to meet specific customer demand. Volume flexibility is the ability to adjust capacity to meet changes in customer quantities. New product flexibility is the ability to launch new or revised products.

Over the past decade, SCM has played an important role for organizations' success, and subsequently, the green supply chain (GSC) has emerged as an important component of the environmental and supply chain strategies of a large number of companies. According to Gupta, environmental management relieves environmental destruction and improves environmental performance by institutionalizing various greening practices and initiating new measures and developing technologies, processes, and products.

Some studies focused on external environmental factors such as customers and suppliers. To improve their own environmental supply chain performance, organizations need the interactions with the government, suppliers, customers, and even competitors (Carter and Ellram 1998). Cooperation with suppliers and customers has become extremely critical for the organizations to close the supply chain loop (Zhu et al. 2008). Moreover, Jayaraman et al. (2012) suggest that adopting an environmental perspective on operations can lead to improved operations. The study further suggests that any operational system that has minimized inefficiencies is also more environmentally sustainable.

On the contrary, Bouchery et al. (2012) identified the need of study of quantitative models in operations management. Their study contributes by revisiting classical inventory methods taking sustainability concerns into account. They designed sustainable order quantity model which can help decision makers to quickly identify the best option among these solutions. Similarly, Dinwoodie et al. (2012) designed a framework to facilitate environmental management applies business process principles to identify relevant inputs, processes, and outputs. The study further suggests that simplification and optimization phases of business process re-engineering can be tapped by business strategists for enhancing adoption of sustainable practices

(Acar et al. 2010). Tang and Zhou suggest the correlation between profitability and sustainability. However, the balance can only be maintained in the long run if the firm can take a holistic approach to sustain the financial flow (profit), resource flow (planet), and development flow (people) for the entire ecosystem comprising poor producers in emerging/developing markets, global supply chain partners, consumers in developed countries, and the planet (Ageron et al. 2012; Rastogi et al. 2013).

Likewise, Sarkis (2012) studied the social issues such as sustainability, poverty alleviation, health care management, philanthropic activities, humanitarian aid, and education can all benefit from modeling efforts from operations management and production economics researchers (Goel and Jindal 2013). The author defined compassionate operations and included the culmination of major natural and man-made crises, increased environmental concerns, and increased globalization and knowledge, comes a wider awareness of social problems that need to be addressed (Walsh and Dowding 2012).

Dey et al. (2011) examined the current state of sustainability efforts within the field of supply chain management, more specifically supply chain logistics operations, and to identify opportunities and provide recommendations for firms to follow sustainable operations. The study shows that for firms to implement a sustainability strategy in their supply chain operations, the logistics function needs to play a prominent role because of the magnitude of costs involved and the opportunity to identify and eliminate inefficiencies and reduce the carbon footprint (Dey et al. 2011). Huang et al. suggested the concept of closed-loop supply chain (CLSC) management to meet the current environmental challenges and sustainable development. The study suggests that mainly three challenges are there for SCM, namely (1) uncertainty of time delay in remanufacturing and returns, (2) uncertainty of system cost parameters, and (3) uncertainty of customers' demand disturbances. The study was based on operations of scrap supply chain in the Chinese steel industry. Ageron et al. (2012) study suggests that sustainability research on supply management has received limited attention. Similarly, Schoenherr (2012) investigated the influence of sustainable business development on manufacturing plant operations focusing on the environmental component. Specifically, on the basis of resource-based view of the firm, the authors have hypothesized the impact of environmental management on plant performance. However, their ability to realize sustainable competitive advantage is hampered by the lack of sustainable objective.

## 3   Research Objective

The present study is aimed at

- To study the environmental criteria for the section of suppliers in the SCM
- To identify the unique factor for the reduction in waste to be environmentally sustainable.

**Table 1** Result of KMO and Bartlett's test

| Kaiser-Meyer-Olkin Measure of sampling adequacy | | 0.777 |
|---|---|---|
| Bartlett's test of sphericity | Approx. Chi-square | 101.320 |
| | df | 10 |
| | Sig. | 0.000 |

## A. Methodology

This paper included the respondents from both manufacturing and service firms located in India. The sampling companies which have adopted sustainable practices were included in the research. Questionnaires were emailed to top managers of financial sectors. The valid mailing was 520 surveys, from which 450 responses were received. We used only 427 responses. Here, we analyzed 150 data of finance sectors. Therefore, for the study, the response rate is acceptable. Factor analysis has a potential to improve the component loadings. It was used to test the validity of data in the questionnaire. The items were used to measure construct that was extracted to be the only principal component. Table 1 shows the value of Kaiser-Meyer-Olkin Measure of sampling adequacy (KMO) and Bartlett's test of sphericity indicates sustainability of our data for factor analysis. The high value KMO (here KMO = 0.777) indicates that factor analysis may be useful for data analysis. Significance test of Bartlett's test of sphericity gives the result of the test very small value (here 0.000) which indicates that there is a significant relationship present among the variables present. These are statistically significant (Nunnally and Berstein 1994). The scales of the variables were measured also produced internally consistent results. Thus, these measures were found appropriate for analysis. This is because they expressed accepted reliability and validity measures.

## B. Variables

All the variables were measured from the survey. Researchers used a 5-point scale [strongly disagree (1) to strongly agree (5)] to measure all constructs. Sustainable strategy variables were adapted from Lai et al. (2010). These measures assessed the degree of practice, processes, and decision-making activities which help to implement the sustainable strategy. Measures of operation flexibility were taken for both centrality and complexity (Adamides and Karacapilidis 2005).

# 4 Result and Analysis

Tables 1, 2, and 3 show the correlation coefficients for all variables. Variance factors (VIFs) were used to check issues relating to multicollinearity, with non-orthogonality in independent variables. The variance ranges from 1.03 to 2.82 well. Hence, there is no multicollinearity issues encountered in this study.

**Table 2** Total variance explained

| Component | Initial eigenvalues | | | Extraction sums of squared loadings | | |
|---|---|---|---|---|---|---|
| | Total | % of variance | Cumulative (%) | Total | % of variance | Cumulative (%) |
| 1 | 2.638 | 52.769 | 52.769 | 2.638 | 52.769 | 52.769 |
| 2 | 0.873 | 17.451 | 70.219 | 0.873 | 17.451 | 70.219 |
| 3 | 0.584 | 11.673 | 81.893 | | | |
| 4 | 0.513 | 10.251 | 92.143 | | | |
| 5 | 0.393 | 7.857 | 100.000 | | | |

Extraction method: principal component analysis

**Table 3** Component matrix

| | Component | |
|---|---|---|
| | 1 | 2 |
| BP405 | 0.818 | 0.238 |
| BP401 | 0.759 | $-4.71E-02$ |
| BP403 | 0.749 | $-0.241$ |
| BP402 | 0.655 | $-0.588$ |
| BP404 | 0.635 | 0.640 |

Extraction method: principal component analysis is a two-component extracted

**KMO and Bartlett's Test**

## 4.1 Reliability Assessment

Internal consistency reliability measures were assessed on the factor structures derived from both analyses reported above, using Chronbach's procedure available in the SPSS statistical package. Those variables have been retained for which the values of Chronbach's alpha are more than 0.5.

| Variables | Cronbach's alpha |
|---|---|
| Green policy | |
| Green technology | |
| Green human resource management | |
| Green supply chain management | |
| Green production flexibility | |

It suggests that having clear plans also influences the natural environment training programs for managers and employees which in turn enhance the performance of sustainability strategy. Moreover, the first major predictor of sustainability is designing environmental policy (EP) as shown in table. The other predictors are natural environment training programs for managers and employees. The corresponding ANOVA values for the regression model are shown here as indicating validation at 99% confidence level. The coefficient summary gives beta values of environmental policy (EP), training the employees on sustainability (TR), and sustainability outcomes measured as customer preference for green products (GP), which are fairly representative of their impact on sustainability. Thus, environmental policy (EP) is emerging as a major influence variable for sustainability.

# References

Acar, Y., Kadipasaoglu, S., & Schipperijn, P. (2010). A decision support framework for global supply chain modelling: An assessment of the impact of demand, supply and lead-time uncertainties on performance. *International Journal of Production Research, 48*(11), 3245–3268.

Adamides, E., & Karacapilidis, N. (2005). Knowledge management and collaborative model building in the strategy development process. *Knowledge and Process Management, 12,* 77–88. https://doi.org/10.1002/kpm.223.

Ageron, B., Gunasekaran, A., & Spalanzani, A. (2012). Sustainable supply management: An empirical study. *International Journal of Production Economics*, 140. https://doi.org/10.1016/j.ijpe.2011.04.007.

Al-Fawaz, K., Al-Salti, Z., & Eldabi, T. (2008). Critical success factors in ERP implementation: A review. *European and Mediterranean Conference on Information Systems 2008 (EMCIS2008)* May 25–26.

Amit, R., & Zott, C. (2010, July). *Business model innovations: Creating value in times of change, working paper-870.*

Bouchery, Y., Ghaffari, A., Jemai, Z., & Dallery, Y. (2012). Including sustainability criteria into inventory models. *European Journal of Operational Research, 222,* 229–240. https://doi.org/10.1016/j.ejor.2012.05.004.

Carter, C. R., & Ellram, L. M. (1998). Reverse logistics: A review of the literature and framework for future investigation. *Journal of Business Logistics.* ISSN: 0197-6729, EISSN: 2158-1592.

Davis, F. D. (1985). A technology acceptance model for empirically testing new end-user information systems.

Dinwoodie, J., Tuck, S., & Knowels, H. (2012). Assessing the environmental impact of maritime operations in ports: A systems approach. In *Maritime logistics: Contemporary issues* (pp. 263–283).

Dyer, L., & Ericksen, J. (2009). Complexity-based agile enterprises: putting self-organizing emergence to work. In A. Wilkinson, et al. (Eds.), *The sage handbook of human resource management* (pp. 436–457). London: Sage.

Gavronski, I., Klassen, R., Vachon, S., & Nascimento, L. (2011). A resource-based view of green supply management. *Transportation Research Part E-Logistics and Transportation Review, 47,* 872–885. https://doi.org/10.1016/j.tre.2011.05.018.

Gligor, D. M., Esmark, C. L., & Holcomb, M. C. (2015). Performance outcomes of supply chain agility: When should you be agile? *Journal of Operations Management, 33–34,* 71–82.

Goel, N., & Jindal, A. (2013). Implementation of cost effective solution for eGovernance through APUS (Aadhar Card, PPP, Underprivileged, SMS). In *Proceedings of the 7th National Conference, INDIACom-2013*.

Holland, C. P., & Light, B. (1999, January). A critical success factors model for enterprise resource planning implementation, Conference Paper.

Hollos, D., Blome, C., & Foerstl, K. (2012). Does sustainable supplier cooperation affect performance? Examining implications for the triple bottom line. *International Journal of Production Research, 50*(11), 2968–2986.

Jayaraman, K., Kee, T. L., & Soh, K. L. (2012). The perceptions and perspectives of Lean Six Sigma (LSS) practitioners. *The TQM Journal, 24*(5), 433–446.

Jonás Montilva, C., & Judith Barrios, A. (2004, December). BMM: A business modeling method for information systems development. *Clei Electronic Journal, 7*(2), PAPER 3 (2004).

Ketchen, D., Boyd, B., & Bergh, D. (2007). Research methodology in strategic management: Past accomplishments and future challenges. *Organizational Research Methods, 11*, 643–658. https://doi.org/10.1177/1094428108319843.

Lai, C.-S., Chiu, C.-J., Yang, C.-F., & Pai, D.-C. (2010). The effects of corporate social responsibility on brand performance: The mediating effect of industrial brand equity and corporate reputation. *Journal of Business Ethics, 95*, 457–469. https://doi.org/10.1007/s10551-010-0433-1.

Leybourn E. (2013). Directing the agile organisation: A lean approach to business management.

Nah, F. F., Tan, X., & Teh, S. H. (2004). An empirical investigation on end-users' acceptance of enterprise systems. *Information Resources Management Journal, 17*(3), 32–53.

Nunnally, J. C., & Bernstein, I. H. (1994). The capability approach and evaluation of the well-being in Senegal: An operationalization with the structural equations models, the assessment of reliability. *Psychometric Theory, 3*, 248–292.

Qrunfleh, S., & Tarafdar, M. (2013). Lean and agile supply chain strategies and supply chain responsiveness: The role of strategic supplier partnership and postponement. *Supply Chain Management: An International Journal, 18*, 571–582.

Rastogi, R., Ilango, T., & Chandra, S. (2013). Pedestrian flow characteristics for different pedestrian facilities and situations.

Report of the World Commission on Environment and Development. Our common future (WECD'1987), pp. 1–247.

Sarkis, J. (2012). A boundaries and flows perspective of green supply chain management. *Supply Chain Management, 17*, 202–216. https://doi.org/10.1108/13598541211212924.

Schoenherr, T. (2012). The role of environmental management in sustainable business development: A multi-country investigation. *International Journal of Production Economics, 140*. https://doi.org/10.1016/j.ijpe.2011.04.009.

Seebach, C., Beck, R., & Denisova, O. (2013). Analyzing Social Media for Corporate reputation management: How firms can improve business agility. *International Journal of Business Intelligence Research (IJBIR), 4*(3).

Shafer, S. M., Smitha, H. J., & Linder, J. C. (2005). The power of business models. *Business Horizons, 48*, 199–207.

Tse, Y.K., Zhang, M., Akhtar, P., & MacBryde, F. (2011). Embracing supply chain agility: An investigation in the electronics industry. *Journal of Supply Chain Management, 21*(1), 140–156.

Tseng, Y. H., & Lin C. T. (2016). Enhancing enterprise agility by deploying agile drivers, capabilities and providers. *Information Sciences, 181*, 3693–3708.

Vickery, S., Canlantone, R., & Droge, C. (1999). Supply chain flexibility: An empirical study. *Journal of Supply Chain Management, 35*(1), 16–24.

Walsh, H., & Dowding, T. J. (2012). Sustainability and the Coca-Cola company: The global water crisis and Coca-Cola's business case for water stewardship. *International Journal of Business Insights & Transformation, Special Issue, 4*, 106–118.

Zhu, Q., & Sarkis, J. (2004). Relationships between operational practices and performance among early adopters of green supply chain management oractices in Chinese manufacturing enterprises. *Journal of Operations Management, 22*, 265–289. https://doi.org/10.1016/j.jom.2004.01.005.

Zhu, Q., Sarkis, J., & Lai, K. (2008). Green supply chain management implications for closing the loop. *Journal of Transportation Research Part E, 44*, 1–18.

# Waste Recycling in a Developing Context: Economic Implications of an EU-Separate Collection Scheme

Amani Maalouf, Francesco Di Maria and Mutasem El-Fadel

**Abstract** This study assesses the economic viability of implementing a successful developed economy-based separate collection scheme in a developing economy test area while taking into consideration different influential factors. Two scenarios with different intensities of source segregated (SS) materials were simulated to compare the overall collection cost in developing versus developed economies while considering the variation in waste composition. The SS efficiencies were calculated based on a successful source separation scheme implemented in a developed economy. Scenario S1 reflects a policy towards separation of paper and packaging waste with an overall SS intensity of 13% in the test area in comparison with 25% in the developed economy. Scenario S2 considered an increase in the overall SS intensity that reached 68% in the test area in comparison with 48% in developed economy, when considering the separation of organic waste. The results showed that in the test area, an increase in SS intensity from 13% up to 68% caused a significant reduction in residual municipal solid waste but a consequent increase in the overall collection cost reaching up to ~44%. The developing economy exhibited significantly lower (63–84%) collection costs in comparison with developed economy, mainly due to significantly lower personnel cost. Variation in waste composition caused a major difference in the overall collection cost between developing and developed economies, depending on waste density, collection vehicles load, and compaction ratio. For instance, the collection of low-density waste (e.g. light packaging) resulted in lower fuel consumption and collection cost (up to 83%) in developing economies in comparison with higher fractions in developed economies.

**Keywords** Waste separation · Waste collection · 3R concept · Economic assessment · Developing context

A. Maalouf (✉) · M. El-Fadel
Department of Civil and Environmental Engineering, American University of Beirut, Riad El-Solh, Beirut 1107 2020, Lebanon
e-mail: ahm22@mail.aub.edu

F. Di Maria
LAR Laboratory, Department of Engineering, University of Perugia, Via G. Duranti, 06125 Perugia, Italy

CRIC Consortium, Jadavpur University, Kolkata, India

© Springer Nature Singapore Pte Ltd. 2020
S. K. Ghosh (ed.), *Waste Management as Economic Industry Towards Circular Economy*,
https://doi.org/10.1007/978-981-15-1620-7_5_11

# 1   Introduction

Worldwide, nearly more than 2 billion tonnes of solid waste are generated annually, with projections to reach 3.4 billion tonnes by 2050 (Kaza et al. 2018). Population growth, development, and urbanization have increased the quantities of municipal solid waste generation to levels raising considerable management challenges (Gundupalli et al. 2017). The 3R (reuse, recycle, and recovery) concept has been evolving to become most effective in partially facing these challenges. This concept is extrapolated from the waste management hierarchy, which was developed by the European Commission and was long recognized in the EU legislation as a fundamental component of integrated waste management (Council Directive 1991). In this context, separation of waste material at source is a critical factor influencing the successful implementation of this concept. Several studies (Boonrod et al. 2015; Sukholthaman and Sharp 2016) have also demonstrated the effectiveness of waste separation at source in reducing the amount of waste to be landfilled and increasing the amount of recyclable materials. Accordingly, it has been widely applied in developed economies (Rousta et al. 2015; Di Maria and Micale 2014) towards a sustainable integrated waste management system. Developing economies have witnessed a lack of public participation in waste separation at source with limited applications in pilot cities (Kaza et al. 2018; Tai et al. 2011). This can be attributed to several factors such as the lack of awareness about the importance of waste separation at source (Kaza et al. 2018; Boonrod et al. 2015), outdated legislation or lack of services and infrastructure (Sukholthaman et al. 2017), unavailability of market for recyclables (Belton et al. 1994), and inconsistent waste separation campaigns (Miller Associates 1999). In turn, waste collection of source-separated material can affect its quality and consequently can impact the effectiveness and efficiency of the 3R process. The waste collection process in developing economies shares the highest cost among the other urban services whereby local authorities spend between 20 and 50% of their budget on this service (UN Habitat 2010).

Past efforts evaluated the impact of source separation on waste collection and identified influencing factors affecting the application of this concept (Sukholthaman and Sharp 2016; Vassanadumrongdee and Kittipongvises 2018; Boonrod et al. 2015). Other studies examined aspects related to source segregation intensity, fuel consumption, as well as economic and environmental impacts (De Oliveiera and Borenstein 2007; Di Maria et al. 2013; Everett et al. 1998a, b; Iriarte et al. 2009; Johansson 2006). For instance, Sukholthaman and Sharp (2016) demonstrated that the higher source separation rate, the less amount of waste left to be landfilled, the less the total management cost waste to be collected, and eventually the higher the efficiency of the waste collection service. However, the authors did not consider additional costs from collecting source-separated waste.

While economic implications of implementing separate waste collection schemes have been recognized particularly in developed economies (Di Maria and Micale 2013), limited to no study identified the differentiating factors influencing the practical implementation of this concept in a developing context. These factors may include the amount and quality of the waste, its composition and fraction of recyclable materials, available recycling industries and market for recyclable material, public awareness and attitude, as well as the economic, legal, and institutional support to the 3R concept. In this study, we aim to assess the economic viability of implementing a successful developed economy-based separate collection scheme in a developing economy context, while considering those influential factors. The ultimate objective is to support the development of an economically viable separate collection system while quantifying advantages and disadvantages towards decision making and policy planning.

## 2 Materials and Methods

### 2.1 Test Area

The test area (Beirut, Lebanon) has a population >2 M inhabitants and encompasses mostly medium to high-rise apartment buildings. Table 1 presents its average waste composition in comparison with a medium-sized Italian city with characteristics similar to the test area. Generally, developing economies are characterized by a higher fraction of organic waste (53.4%) in comparison with developed economies (20.3%). Papers (15.6%) in developing economies are lower than those encountered in developed economies of 35.5% (Table 1).

The management system in the test area consists of commingled MSW collection, sorting and recycling, composting, and landfilling. Waste is collected at a cost of 26 €/tonne (CDR 2010) and transferred into two material recovery facilities (MRFs) where it is sorted into bulky items, inerts, biodegradable organics, and recyclables. The biodegradable fraction is sent for open windrow composting with relatively

**Table 1** Average MSW composition (% w/w)

| Waste category | Test area[a] | Developed economy[b] |
|---|---|---|
| Organic | 53.4 | 20.3 |
| Glass | 3.4 | 7 |
| Metals | 2 | 6.5 |
| Papers | 15.6 | 35.5 |
| Plastics | 13.8 | 12.6 |
| Textiles | 2.8 | 1.5 |
| Wood | 0.8 | 3.6 |
| Others | 8.2 | 12.7 |
| Total | 100 | 100 |

[a]Laceco/Ramboll (2012); Maalouf and El-Fadel (2019a)
[b]Data retrieved from Di Maria and Micale (2013) for a typical Italian city

low-quality compost often rejected by farmers (Maalouf and El-Fadel 2019b). Waste management activities concerning the collected waste after being unloaded at the MRFs were not been included in the present study.

## 2.2 Scenario Definition: Policy Management and Economic Analysis

In this study, a simulation model (Di Maria and Micale 2013) was used to calculate associated costs of adopting a separate waste collection system. The model runs giving to space/time correlation and is able to estimate the quantity of collected waste (tonnes), the amount of fuel consumed (L), and the time (s) required to cover a given collection route (km). Input data are presented in Table 2.

Two scenarios with different intensities of source segregated (SS) materials were simulated to compare overall collection cost with respect to a developed economy-based separate collection scheme while considering the difference in waste composition (Table 3). The SS efficiencies were calculated based on a successful source separation scheme implemented in a developed economy (Di Maria and Micale 2013). The segregation efficiency by individual waste component adopted for all scenarios is displayed in Table 4 Scenario S1 reflects a policy towards separation of paper and packaging waste with an overall SS intensity of 13% in the test area in comparison with 25% in a developed economy. The latter is about double the overall SS intensity in the test area due to the higher fraction of recyclable materials (Table 1). An increase in SS intensity was achieved by increasing the amount of source-separated materials. For instance, scenario S2 reached an overall SS intensity of 68% in the test area when considering the organic waste fraction, which is higher

**Table 2** Model input data

| Type of data | Value | Reference |
|---|---|---|
| Vehicle acceleration (km/h/s) | 2.8 | Wang (2001) |
| Pickup time (s) | 60 | Di Maria and Micale (2013) |
| Average speed (km/h) depending on waste collection vehicle (WCV) size (m3) | WCV (22–24 m$^3$): 16<br>WCV (18 m$^3$): 16<br>WCV (6 m$^3$): 16<br>WCV (3 m$^3$): 16 | |
| Average fuel consumption (L/km) depending on WCV size (m$^3$) | WCV (22–24 m$^3$): 0.84<br>WCV (18 m$^3$): 0.70<br>WCV (6 m$^3$): 0.15<br>WCV (3 m$^3$): 0.17 | |
| Vehicle purchase cost (€) depending on WCV size (m$^3$) | WCV (22–24 m$^3$): 18,000<br>WCV (18 m$^3$): 98,000<br>WCV (6 m$^3$): 29,000<br>WCV (3 m$^3$): 18,000 | |

**Table 3** Description of source segregation efficiencies of tested scenarios

| Scenario | Location | Description | Material (% w/w) | Overall SS efficiency (%) |
|---|---|---|---|---|
| S1 | Test area-LB | Separation of paper and packaging (plastic, metals, and glass) | RMSW (86.9) Paper (8.3) Packaging (4.7) | 13 |
|  | Italian city-IT[a] |  | RMSW (74.6) Paper (19.0) Packaging (6.40) | 25 |
| S2 | Test area-LB | Separation of paper, light packaging (plastic), organic, glass, and metals | RMSW (32) Paper (5.1) Light packaging (13.3) Organic (47.1) Glass (1.6) Metals (1) | 68 |
|  | Italian city-IT[a] |  | RMSW (51.9) Paper (11.6) Light packaging (12.1) Organic (17.9) Glass (3.3) Metals (3.2) | 48 |

*RMSW* Residual municipal solid waste; *SS* source segregation; *LB* Lebanon; *IT* Italy
[a]Data retrieved from Di Maria and Micale (2013) for a typical Italian city

**Table 4** Source segregation efficiency by individual waste component for tested scenarios

|  | Waste component | Source segregation efficiency (%) |
|---|---|---|
| S1 | Paper | 53.23 |
|  | Light packaging + glass | 24.50 |
| S2 | Paper | 32.70 |
|  | Light packaging | 96.00 |
|  | Organic | 88.20 |
|  | Glass | 47.10 |
|  | Metal | 49.20 |

than at the developed economy (Table 1). Therefore, the latter resulted in a lower SS intensity of 48% in S2 (Table 3). The waste collection vehicles (WCV) are equipped with rear loaders and powered by diesel fuel oil at an average cost of about 0.55 €/L. Larger WCV (i.e. 18–24 m$^3$) operate with a 2-person crew, while 6 and 3 m$^3$ WCV operate with 1-person crew.

Data retrieved from Di Maria and Micale (2013) for a typical Italian city

In all scenarios, the simulations used similar assumptions for a separate collection scheme in a developed economy, which include:

- Maximum length of a work shift is 6 h per day;
- Two work shifts per day;
- WCV operating on daily basis with full-day use;
- Vehicle mortgage around 5 years;
- Minimizing number of different size WCV ($m^3$) depending on SS intensity of each scenario. The lower SS intensity scenario S1 requires fewer large-sized WCVs whereby higher SS intensity (S2) requires more small-sized vehicles.

The economic analysis included personnel and vehicle purchasing costs (see Table 2). Vehicle operating and maintenance costs were not considered in this study because they are very similar for the different SS scenarios. The average gross cost per crew member in the test area was assumed at 4500 €/year.

## 3  Results and Discussion

Figure 1 depicts collection costs by waste component for the simulated SS intensity scenarios under developing (LB) versus developed (IT) economies while considering the same amount of waste and variation in waste composition. Collection costs of individual waste components were categorized based on WCV, fuel consumption, and personnel costs.

As expected, the increase in SS intensity up to 68 and 48% in developing and developed economies, respectively, caused a significant reduction in residual MSW (RMSW) but an increase in the collection cost (Fig. 1). For instance, the overall collection cost increased from 9 €/tonne for 13% SS intensity scenario (S1) to 13 €/tonne for 68% SS intensity scenario (S2) (Fig. 1). This can be attributed to the increase in the collection points and total distances travelled on a daily basis by the WCVs. Consequently, this requires an increase in the number of vehicles and personnel involved in the collection activity contributing to the increase in corresponding costs.

Di Maria and Micale (2013) examined the average fuel consumption for 25 and 48% SS intensity scenarios and showed that the fuel consumed per tonne of waste collected increased with SS intensity. The average fuel consumption for the different scenarios ranged from 3.3 to 3.8 L/Ton, respectively, for the 25 and 48% SS intensity. Similarly, the collection costs increased from about 40 to about 70 €/tonne in a typical Italian city characterized by apartment buildings and a high-density population.

Moreover, results showed that for all scenarios, developing economy (LB) resulted in lower collections costs (63–84%) in comparison with developed (IT) economy (Fig. 1). This can be mainly attributed to the personnel cost, which is around 4500 €/year in a developing economy in comparison with 40,000 €/year under

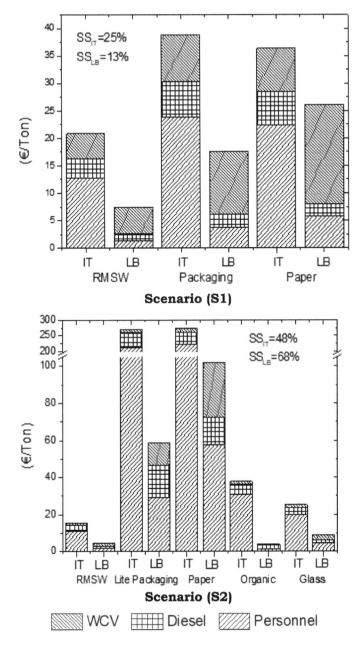

**Fig. 1** Specific collection cost by waste component for different SS intensity scenarios in developed versus developing economies categorized by WCV, diesel, and personnel costs

a developed one. The latter was the major contributor (up to 82%) to the overall collection cost in a developed economy (Fig. 1).

Another aspect that can influence the collection cost is the WCV purchasing cost. In a developing economy, for instance, the WCV purchasing cost was the significant contributor (~67%) to the overall collection cost for the 13% SS scenario (S1), followed by personnel (~20%) and diesel fuel consumption (~13%) costs (Fig. 1). The number of personnel was about 10 at a cost of about 45,000 €/year, operating on two large-sized WCVs (corresponding purchasing cost of about 52,000 €/year). In contrast, the overall collection cost in 68% SS scenario (S2) was influenced mainly by the increase in the number of crew needed (~17 workers at cost of 77,000 €/year) rather than the number of purchased WCVs (one large-sized and five small-sized WCVs for a total of about 58,000 €/year).

Equally important is the impact of waste composition that may vary with location and is noticeably different between developed and developing economies (Table 1). For instance, the collection of low-density waste such as light packaging and RMSW resulted in higher fuel consumption and collection cost in developed economies in comparison with lower fractions under a developing economy (Fig. 1). In particular, the collection of low-density waste performed in high SS intensity scenario (S2) resulted in high collection costs due to the low weight transported by the WCVs. In contrast, the collection of source segregated organic waste in scenario S2 resulted in lower collection cost in comparison with other waste components as a consequence of the lower fuel consumption, which is greatly affected by the lower compaction ratio of the large-sized WCVs with respect to the low-sized WCVs in S1.

# 4  Conclusion

Waste collection of source-separated waste is an essential component of an integrated waste management system whereby it can affect its quality and consequently can impact the effectiveness and efficiency of the 3R process. This study assessed the economic viability of implementing a successful developed economy-based separate collection scheme in a developing economy test area while taking into consideration different influential factors and variation in waste composition. The results showed that the increase in SS intensity from 13% up to 68% caused a significant reduction in residual municipal solid waste but a consequent increase in the overall collection cost reaching up to ~44%. Note that for all tested scenarios, developing economy exhibited lower collections costs (63–84%) in comparison with developed economy, mainly due to significantly lower personnel cost. Moreover, a comparison of average reported data showed that the collection cost is affected mainly by the increase in purchasing waste collection vehicles and personnel costs rather than fuel consumption cost. Differences in waste composition between developed and developing economies also played a significant role in affecting the overall collection cost, depending on waste density, collection vehicles load, and compaction ratio. For instance, the collection of low-density waste (e.g. light packaging) resulted in lower fuel consumption and collection cost (up to 83%) in developing economies in comparison with higher fractions in developed economies.

**Acknowledgements** This study is supported through joint funding from the National Council for Scientific Research (NCSR) and American University of Beirut (AUB). Special thanks are extended to Dar Al-Handasah (Shair & Partners) for its support to the graduate programs in Engineering at AUB.

# References

Belton, V., Crowe, D. V., Matthews, R., & Scott, S. (1994). A survey of public attitudes to recycling in Glasgow. *Waste Management and Research, 12,* 351–367.

Boonrod, K., Towprayoona, S., Bonneta, S., & Tripetchkul, S. (2015). Enhancing organic waste separation at the source behavior: a case study of the application of motivation mechanisms in communities in Thailand. *Resources, Conservation and Recycling, 95,* 77–90.

CDR. (2010). *Progress report on contracts related to sweeping, collection, treatment and disposal of municipal solid waste in Greater Beirut and surroundings (Contract No11707).* Beirut, Lebanon: CDR.

Council Directive. (1991). Directive 91/156/EEC of 18 March 1991 amending directive 75/442/EEC on waste, OJ L 194, 25.7.1975, p. 39 and OJ L 78, 26.3.1991, p. 32.

Di Maria, F., & Micale, C. (2013). Impact of source segregation intensity of solid waste on fuel consumption and collection costs. *Waste Management, 33,* 2170–2176.

Di Maria, F. D., & Micale, C. (2014). A holistic life cycle analysis of waste management scenarios at increasing source segregation intensity: The case of an Italian urban area. *Waste Management, 34*(11), 2382–2392.

Everett, J. W., Maratha, S., Dorairaj, R., & Riley, P. (1998a). Curbside collection of recyclables. I. Route time estimation model. *Resources, Conservation and Recycling, 22,* 177–192.

Everett, J. W., Maratha, S., Dorairaj, R., & Riley, P. (1998b). Curbside collection of recyclables. II. Simulation and economic analysis. *Resources, Conservation and Recycling, 22,* 217–240.

Gundupalli, S. P., Hait, S., & Thakur, A., (2017). A review on automated sorting of source-separated municipal solid waste for recycling. *Waste Management, 60,* 56–74.

Iriarte, A., Gabarrell, X., & Rieradevall, J. (2009). LCA of selective waste collection system in dense urban areas. *Waste Management, 29,* 903–914.

Johansson, O. M. (2006). The effect of dynamic scheduling and routing in a solid waste management system. *Waste Management, 26,* 875–885.

Kaza, S., Yao, L.C., Bhada-Tata, P., & Woerden, F. V. (2018). *What a waste 2.0: A global snapshot of solid waste management to 2050.* Washington, DC: World Bank.

Laceco/Ramboll. (2012). *Preparation of pre-qualification documents and tender documents for solid waste management in lebanon, sub Report 1.* Baseline Study, Beirut, Lebanon: CDR.

Maalouf, A., & El-Fadel, M. (2019a). Life cycle assessment for solid waste management in Lebanon: Economic implications of carbon credit. *Waste Management & Research, 37*(Supplement), 14–26.

Maalouf, A., & El-Fadel, M. (2019b). Towards improving emissions accounting methods in waste management: A proposed framework. *Journal of Cleaner Production, 206,* 197–210.

Miller Associates. (1999). *Project INTEGRA research, attitudes and behaviour, Report 1: Main findings.* Report for the Project Integra Household Waste Research Programme, Hampshire County Council, UK.

Oliveiera, De, Simonetto, E., & Borenstein, D. (2007). A decision support system for the operational planning of solid waste collection. *Waste Management, 27,* 1286–1297.

Rousta, K., Bolton, K., Lundin, M., & Dahlén, L. (2015). Quantitative assessment of distance to collection point and improved sorting information on source separation of household waste. *Waste Management, 40,* 22–30.

Sukholthaman, P., Chanvarasuth, P., & Sharp, A. (2017). Analysis of waste generation variables and people's attitudes towards waste management system: A case of Bangkok, Thailand. *Journal of Material Cycles and Waste Management, 19*(2), 645–656.

Sukholthaman, P., & Sharp, A. (2016). A system dynamics model to evaluate effects of source separation of municipal solid waste management: A case of Bangkok, Thailand. *Waste Management, 52,* 50–61.

Tai, J., Zhang, W., Che, Y., & Feng, D. (2011). Municipal solid waste source-separated collection in China: A comparative analysis. *Waste Management, 31,* 1673–1682.

UN Habitat. (2010). Collection of municipal solid waste in developing countries. *United Nations human settlement programme (UN Habitat), Nairobi, Kenya.*

Vassanadumrongdee, S., & Kittipongvises, S. (2018). Factors influencing source separation intention and willingness to pay for improving waste management in Bangkok. *Thailand. Sustainable Environment Research, 28*(2), 90–99.

Wang, F. S. (2001). Deterministic and stochastic simulations for solid waste collection systems—A SWIM approach. *Environmental Modeling and Assessment, 6,* 249–260.

# A Novel DIY Machine Design to Obtain Secondary Raw Materials from Absorbent Hygiene Waste

**Kunal Gajanan Nandanwar, Divya Rathore and Rajiv Gupta**

**Abstract** Absorbent hygiene waste is unceasingly growing in India. The sanitary napkins and baby/adult diapers, drenched with bodily fluids, are disposed of with regular household wastes and often handled by rag-pickers and waste-collectors with bare hands. These non-compostable wastes are either thrown into sewerage systems, landfills, and water bodies or cremated, posing both serious environmental risks and health issues. This motivates us to design an innovative do-it-yourself (DIY) technological system to recycle used absorbent hygiene products and obtain valuable secondary raw materials in a completely eco-friendly manner. Turning garbage into fuel is potentially an answer to two pressing problems: diminishing the world's dependence on fossil fuels and an alternative to burying trash in landfills. All the major secondary raw materials, 'plastic,' and 'pulp' obtained from the recycling of absorbent hygiene wastes are efficiently converted into useful products, which are demonstrated using an experimental case study.

**Keywords** Absorbent hygiene waste · DIY machine design · Non-compostable wastes

## 1 Introduction

Absorbent hygiene products (AHP) are the products designed to absorb excreted body fluids popularly known as 'diapers,' 'napkins,' or 'sanitary pad' (Kashyap et al. 2016). Life Cycle Assessment of absorbent hygiene waste conducted has found that the largest impact on global warming was caused by the processing of LDPE (low-density polyethylene, a thermoplastic made from the monomer ethylene) used in tampon applicators and the plastic back strip of a sanitary napkin requiring high amounts of fossil fuel-generated energy. A year's worth of a typical feminine hygiene product leaves a carbon footprint of 5.3 kg $CO_2$ equivalents. In the absence of waste collection services, individuals resort to burning, burying, or flushing used pads.[1]

---

[1] https://ecofemme.org/sanitary-waste-in-india/. Retrieved on August 20, 2018.

---

K. G. Nandanwar (✉) · D. Rathore · R. Gupta
Birla Institute of Technology and Science, Pilani, India
e-mail: f2015430@pilani.bits-pilani.ac.in

© Springer Nature Singapore Pte Ltd. 2020
S. K. Ghosh (ed.), *Waste Management as Economic Industry Towards Circular Economy*,
https://10.1007/978-981-15-1620-7_5_12

In the present study, an attempt has been made to provide a comprehensive method for recycling of absorbent hygiene wastes to solve the current waste disposal problems of absorbent hygienic waste. The research aims at (1) reducing the amount of solid waste being disposed into the environment, (2) reclamation of reusable raw materials from disposable waste, (3) biofuel production to save the diminishing reserves of fuel, and (4) recycle the waste through easy to implement the DIY system for real impact on the ecological and economic aspects of the society.

In order to achieve these objectives, an experimental analysis is done to validate the effectiveness of the proposed DIY method. Experimentation is done with various alternatives, and several new techniques are also designed for the process of sterilization, shredding, separation of constituent materials and biofuel production.

## 2   Background Research

### 2.1   Candidate Alternatives for Waste Disposal

#### 2.1.1   Incineration

Incineration is the controlled combustion of waste to generate steam, in turn producing power through steam turbines. According to WHO, incineration of absorbent hygienic waste should be done at temperatures above 800 °C, making it not only difficult to handle such high temperatures but also there is no provision for monitoring the emissions from the incinerator. Incinerators are the leading contributors of dioxin in the global environment.[2]

The burning of plastics, like polyvinyl chloride (PVC), creates highly toxic pollutants. The ultimate release of ash from the incinerator is unavoidable which becomes hazardous waste themselves. The pollutants generated, reside in filters and ash, which require special landfills for disposal. It indirectly fosters continued waste generation while preventing waste prevention, composting, reusing, recycling, and recycling-based economic development.

#### 2.1.2   Burying in a Landfill

The outer layers of AHPs are essentially plastic, which means that they will stay in the landfill for about 800 years. The AHPs in the landfills have a distinctive effect on air pollution, nature, land, and humans. The mixture of toxic substances and decaying organic materials has the compound effects on biodiversity as local vegetation may cease to grow and be permanently altered. As rain falls on landfill sites, organic

---

[2]http://www.alternative-energy-news.info/negative-impacts-waste-to-energy/. Retrieved on August 25, 2018.

**Table 1** General material composition of a commercial sanitary napkin

| Constituent | Composition (%) |
| --- | --- |
| Cellulose wood pulp | 53.44 |
| Plastic top layer | 13.06 |
| Plastic back sheet | 11.40 |
| Super-absorbent polymer gel | 7.13 |
| Silicon paper | 7.95 |
| Hot melt seals | 6.48 |

and inorganic constituents dissolve, forming highly toxic chemicals leaching into groundwater, resulting in serious contamination of the local groundwater.

### 2.1.3 Pyrolysis

Pyrolysis is applied to AHP waste treatment where plastic waste in AHPs is subjected to plastic to the high temperature of $400^-450$ °C in the absence of oxygen, breaking down into smaller molecules of pyrolysis oil, pyrolysis gas, and carbon black. Waste plastic pyrolysis recycles synergy of waste plastic into usable fuel and offers a renewable energy method of waste management.

### 2.1.4 Gasification

Gasification is commonly operated at high temperatures (>600–800 °C) in an air-lean environment (or oxygen-deficient in some applications): The air factor is generally between 20 and 40% of the amount of air needed for the combustion of the plastic solid waste (PSW). Gasification can potentially process both mixed waste and the plastic-only fraction of waste. The process uses a smaller amount of air, resulting in higher energy recovery efficiency and limited formation of pollutants like nitrogen oxides. While gasification is a feasible technology to handle municipal waste, it is commercial applications which have been limited (Brems et al. 2013).

Table 1 shows the general material composition of a commercial sanitary napkin.

## 2.2 DIY Shredding Methods

Various kinds of shredders which produce different shape and size of the shreds were considered.[3]

- **Strip-cut** shredders use rotating knives to cut strips of the same length (or breadth) as the napkins.

---

[3]https://en.wikipedia.org/wiki/Paper_shredder. Retrieved on August 28, 2018.

- **Cross-cut** shredders shred the napkins using contra-rotating drums into parallelogram or diamond-shaped shreds.
- **Cardboard** shredders cut the corrugated material into strips or a mesh pallet.
- **Particle-cut** shredders cut the material into tiny square or circular pieces.
- **Disintegrators** and granulators cut the pulp at random until the particles are fine and small enough to pass through a mesh.
- **Pierce-and-tear** shredders have rotating blades that tear off the napkins.
- **Hammermills** pound the pulp through a screen.
- **Grinders** have cutting blades mounted on a rotating shaft that grind the pulp until it is fine enough to pass through a screen.

The pulp inside the AHP can be shredded using any of the shredders mentioned above, but grinder shredder is experimentally found better due to the ease of shredding and fine quality of chopped pulp.

## 2.3 Sterilization Techniques

The following techniques[4] can be utilized for sterilization of

- **Wet Heat**: The method of autoclaving uses pressurized steam to heat the material to be sterilized. This is highly effective method to kill all microbes, spores, and viruses. Pressurized steam has a high latent heat and holds 7 times more heat than water at 100 °C. This heat is released upon contact with the cooler surface of the material to be sterilized, allowing expeditious delivery of heat and good penetration of dense materials. Sterilization is usually achieved in 15 min at 121 °C by autoclaving.
- **Dry Heat**: Dry heat tends to kill microbes by oxidation of cellular components. This requires more energy than protein hydrolysis, so higher temperatures are required for efficient sterilization by dry heat. Dry heating generally requires a temperature of 160 °C to sterilize.
- **Filtration**: Filters work by passing the solution through a filter with a pore diameter that is too small for microbes to pass through, making it a good way of sterilizing the solutions without heating. For removal of bacteria, filters with an average pore diameter of 0.2 μm are normally used. But virus and phage can pass through these filters making it 'not-reliable' option.
- **Radiation**: UV, X-rays, and gamma rays have intense damaging effects on DNA, making it excellent tools for sterilization. In terms of their effectiveness, the major difference between them is their penetration.
  UV has limited penetration in the air, so sterilization occurs in very limited area near the lamp. X-rays and gamma rays are far more penetrating, which makes them more dangerous but very effective for large-scale cold sterilization of plastic items during manufacturing.

---

[4]https://bitesizebio.com/853/5-laboratory-sterilisation-methods/. Retrieved on July 24, 2018.

To process an easy to implement and safe sterilization, autoclaving was chosen. Autoclaving requires lower temperature in comparison with dry heating and could be economically more effective in comparison with filtration and radiation techniques.

## 2.4 Biofuel Production from Plastics

Pyrolysis of waste plastics into fuel is one of the best means of conserving valuable petroleum resources in addition to protecting the environment. This process involves catalytic degradation of waste plastic into fuel range hydrocarbon, i.e., petrol, diesel, and kerosene, etc. It is a catalytic cracking process in which waste plastic is cracked at very high temperature and the resulting gases are condensed to recover liquid fuels. Type of plastics also affects the rate of conversion of fuel, and the results of this process are found to be better than other alternate methods which are used for the disposal of waste plastic (Joshi and Rambir 2013).

Recycling of absorbent hygiene waste is not a common practice. The existing methods of recycling absorbent hygiene products (AHPs) use sophisticated engineering and machines which generally do not have good techno-economic viability. The first-of-its-kind DIY segregation of degradable and non-biodegradable substances discussed in the paper is a state-of-the-art concept which can solve the critical issue of waste management of billions of tons of AHPs produced each year and can produce novel by-products as well. The system can be established in a community with 15–20 houses. The designed machine is semi-automatic, and the products received after separation are utilized efficiently to form finished products. The machine is made with various readily available components and can be operated by an unskilled worker as well. If the finished products are sold effectively, then the revenue generated can make the system self-sustainable as well.

## 3 DIY Mechanical Design

Figure 1 shows the main phases of our designed DIY recycling process for AHPs.

## 3.1 Shredding

Shredding is done to convert the collected waste napkins into smaller pieces for sterilization and separation. We have chosen a grinder system as shredder and experimentally tested the grinding of soiled and fresh napkins. A total of 186 grams of used absorbent hygiene products were put in the grinder jar of 750 ml capacity. Grinder system converts the pulp to pieces of mm size, whereas the plastic top and bottom layers are not shredded to fine pieces but get detached from the pulp after

**Fig. 1** Flowchart of the
processes involved in DIY
recycling of AHPs

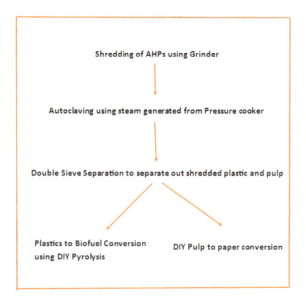

grinding. This difference in the size of shredded paper and plastic components serves
as the basic principle for separation after shredding. Figures 2 and 3 show the grinder
assembly used and mixture after grinding, respectively.

**Fig. 2** Grinder system used
to experiment with 750-ml
jar volume and 186 grams of
absorbent hygiene waste

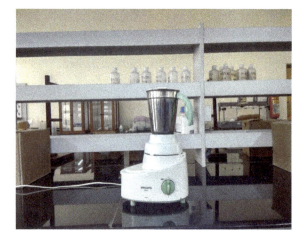

**Fig. 3** Shredded waste in the grinder jar

## 3.2  Sterilization

The mixture of shredded AHPs is transferred from grinder to the top sieve of the double sieve assembly. Autoclaving performed using a pressure cooker is chosen to sterilize the waste. Water is heated in a pressure cooker using parabolic reflectors on the sunny days. An electric heating system is there to heat the water whenever solar radiation is not available. If electric heaters are not feasible, then a gas cylinder system can also be accommodated in the design as the container storing the heating system has holes for air inlet and exhaust.

A pipe is connected to the pressure cooker weight valve which passes the steam to the upper sieve where shredded waste is present. At 15 PSI, the pressure cooker produces steam at 121 °C. At this temperature and pressure, sterilization for about 15 min kills most of the microorganisms in the AHPs. The steam is passed to the waste mixture by a perforated pipe to sterilize the waste effectively (French and Greenville 1942).

## 3.3  Double Sieve Separation

Double sieve separation designed by us proves to be the best alternative to separate plastic and pulp waste from shredded absorbent hygienic waste. Figures 4 and 5 show the simple apparatus of the double sieve which can separate the plastic and pulp as they have different sizes. This separation technique does not require any power like centrifuge separation, etc., and at the same time, it is faster than the sedimentation technique generally used for such kind of separation.

Various combinations of sieves were experimentally checked. Two sieves of sizes 10 and 1.18 mm were chosen with 10 mm sieve as the top sieve, although various

**Fig. 4** Double sieve
separation experimental
setup using 10 mm top sieve
and 1.18 mm bottom sieve

**Fig. 5** Separated plastic in
top sieve and pulp in bottom
sieve

sieves between 9 to 12 mm and 1.1 mm to 2 mm, respectively, can be used depending upon their availability in the market. Water containing alum is continuously flown over the sieves which drive the waste to pass through sieves, with alum removing the super-absorbent polymer (SAP) from the grinded waste (Grimes 2012). The top sieve allows everything except shredded plastic as shredded plastic has a bigger size after grinding. The lower sieve of 1.18 mm size allows only the water to pass through, and pulp is retained in the sieve. The collected plastic and pulp in the different sieves are then used to make biofuel and paper, respectively. The flowing water is recirculated from bottom of the container to top sieve using commonly available water pump used in desert coolers.

The completely DIY assembly shown in Figs. 6 and 7 for the segregation of plastic and pulp is designed in a compact form with two closed chambers. One chamber has grinder assembly and pressure cooker system. The shredded waste after grinding is manually transferred to top sieve in chamber 2 using the grinding jar. Steam

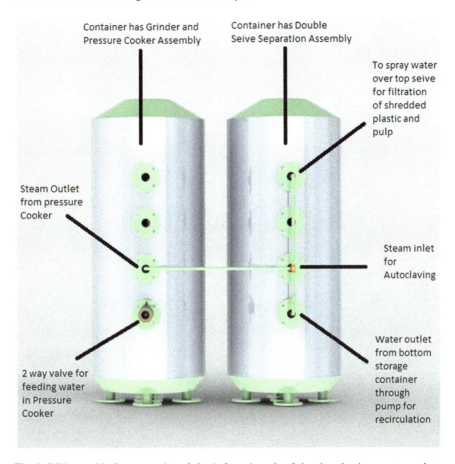

**Fig. 6** DIY assembly for segregation of plastic from the pulp of absorbent hygiene waste products

generated in chamber 1 through the pressure cooker goes to chamber 2 and passes over the grinded waste kept in the top sieve from bottom to top. When the sterilization is done and waste is cooled for about an hour after sterilization, then the recirculation pump is started which sprays water through a connected nozzle over the sieves to drive out filtration in the sieves. The used up water is replenished from the top valve in chamber 2. When all the plastic waste is collected in the top sieve and all the pulp waste is retained in bottom sieve, the segregated waste can be collected from assembly. Plastic is sent for DIY biofuel production, and paper is made from pulp.

Little space is kept below the pressure cooker to accommodate the gas burner if gas-based pressure cookers are used for steam generation. Also, multiple tube openings are provided in both the chambers to accommodate the different sized components as per their availability in the market. All these factors have been thought of while designing this DIY system, thereby making it a multifarious system.

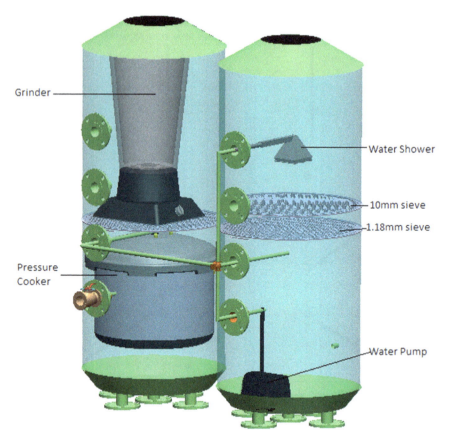

**Fig. 7** DIY assembly for segregation of plastic from pulp of absorbent hygiene waste products with inner components labeled

## 3.4 Plastic to Biofuel

Pyrolysis of plastics gives valuable fuels like gasoline, kerosene, and diesel. One kg of fuel yields up to 1 kg of fuel. This process needs a DIY pyrolysis reactor which is an airtight vessel attached with a condensation tube made up of copper. A thermocouple also needs to be attached to monitor the temperature. In the reactor, the plastic is converted to vaporized fuel above 400 °C. The vapor produced is condensed to liquid by passing it through a condenser made up of copper or steel. The vapor at 400 °C requires a longer length of the tube to cool down, so the vapors are bubbled into water to cool them down. Figure 8 shows the components of the pyrolysis of the plastic process.

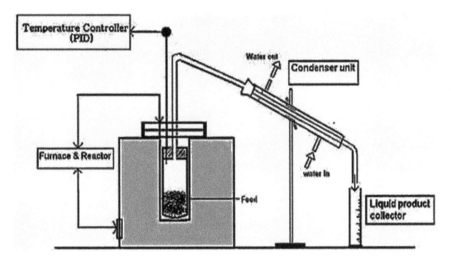

**Fig. 8** Setup for plastic pyrolysis to convert plastics to biofuel (http://www.gallactronics.com/2013/11/how-to-convert-plastic-to-oil.html. Retrieved on August 26, 2018.)

## 3.5 Secondary Products from Pulp

There are techniques to convert pulp to biofuel, but the process is not economically viable so the pulp produced in the present DIY machine can be put to other various uses like papermaking and recycled finished products. Pulp from several machines can be collected together for processing it to make paper.

## 4 Cost Analysis and Comparison

Several waste management companies like Knowaste Inc., Fater, etc. do recycling of AHPs in big plants with setup and operational costs of thousands of dollars (Gerina-Ancane and Eiduka 2016). Such unfeasible costs make it impossible to be set up by government or municipality of the country. However, our designed system is very compact which can be installed similar to municipality dustbins in every community as shown in Fig. 9.

Table 2 shows the approximate cost analysis of the designed DIY system. The setup cost of a unit is about $200 which makes it feasible to set up in communities. The operation cost of the plant can be born by the produced by-products, therefore, making the system sustainable.

**Fig. 9** View of our designed
DIY system in rural areas

**Table 2** Approximate cost
analysis of one unit of DIY
recycler

| Component used in designed machine | Cost |
|---|---|
| 2-L grinder system | $42 |
| 3-L pressure cooker | $7 |
| Sieves | 21 |
| Pipes and valves | $25 |
| Water heating system | $57 |
| Pyrolysis reactor and condensation tube | $14 |
| Outer body for safety and other attachments | $22 |

## 5 Results and Conclusions

The DIY technological design represents an absolute novelty compared with the
alternatives currently in operation. None of the phases entails human contact with
AHP waste, so it is a major breakthrough to improve hygiene practices throughout
the world. The secondary raw materials produced prove to be higher added value
products and strengthen the economic sustainability of the project as a whole.

**Acknowledgements** We wish to acknowledge Dr. Prabhat N. Jha—Associate Professor and Head of Department of Biological Sciences, BITS Pilani, Pilani, Mr. Ajay K. Yadav—Laboratory Assistant, Microbiology Laboratory, BITS Pilani, Pilani, and Mr. Rais Dayal Soni—Laboratory Assistant, Geotechnical Laboratory, BITS Pilani, Pilani, for their invaluable discussion, guidance, and support throughout the course of this project.

# References

Brems, A., Dewil, R., Baeyens, J., & Zhang, R. (2013). Gasification of plastic waste as waste-to-energy or waste-to-syngas recovery route. *Natural Science, 05*(06), 695–704. https://doi.org/10.4236/ns.2013.56086.

French, C. A., & Greenville, S. C. (1942). Sterilizing apparatus and process. *United States Patent Office, 2*(388), 876.

Gerina-Ancane, A., & Eiduka, A. (2016). Research and analysis of absorbent hygiene product (AHP) recycling. In *15th International Scientific Conference: Engineering for Rural Development*, ISSN 1691-5976.

Grimes, D. (2012). Separation of materials comprising super absorbent polymers using reduced water. *United States Patent, US, 2012*(0217326), A1.

Joshi, A., & Rambir, P. R., (2013, November). *Conversion of plastic wastes into liquid fuels— A review, recent advances in bioenergy research* (3rd ed.). https://www.researchgate.net/publication/281064326_CONVERSION_OF_PLASTIC_WASTES_INTO_LIQUID_FUELS_-_A_REVIEW.

Kashyap, P., Win, T. K., & Visvanathan, C. (2016). Absorbent hygiene products—An emerging urban waste management issue. In *Asia-Pacific Conference on Biotechnology for Waste Conversation.*

# Sustainable Waste Transportation in Kolkata Using DBMS and Image Sensing

**Rahul Baidya, Ipsita Saha, Tapobrata Bhattacharya and Sadhan Kumar Ghosh**

**Abstract** The purpose of this study is to generate a robust waste transportation system for the city of Kolkata to the proposed waste-to-energy (WTE) plant and to the existing composting plant. WTE plants are concerned with generating power thus inorganic fragments is the primary concern, on the other hand composting plants primarily needs organic fragments. In this study, a database management system (DBMS) was developed for selecting the compactor waste which will be towed to the truck to be carried to the WTE plant or composting plant. The waste-to-energy plant will be able to identify the waste composition coming to the plant, thus making the supply chain more holistic and robust. The parameter was identified from literature review, and a field study was carried out to identify the composition of wastes dumped into compactor in different areas. The composition of the waste is then gauged by the image sensing process. The algorithm will have a predefined value based on which the truck destination will be decided. The data of different trucks carrying waste will be stored in the DBMS. The system will reduce the transportation cost considerably by making the decision as soon as the waste arrives at the compactor. The results if implemented could reduce the cycle time and could provide an effective solution to analyze the waste before it is transported to the plants. The DBMS-image sensing system (DBMS-ISS) will analyze the surface characteristics of the waste. Thus, this process will generate the approximate composition of the waste.

**Keywords** Waste · DBMS · Image sensing · Kolkata

R. Baidya (✉) · I. Saha · T. Bhattacharya
Department of Mechanical Engineering, Jadavpur University, Kolkata, India
e-mail: rahulbaidya.ju@gmail.com

R. Baidya · T. Bhattacharya
Centre for Research and Innovation, Department of Mechanical Engineering, Institute of Engineering & Management, Kolkata, India

I. Saha · S. K. Ghosh
Department of Computer Science and Engineering, Guru Nanak Institute of Technology, Panihati, Sodepur, Kolkata, India

© Springer Nature Singapore Pte Ltd. 2020                                      129
S. K. Ghosh (ed.), *Waste Management as Economic Industry Towards Circular Economy*,
https://10.1007/978-981-15-1620-7_5_13

# 1 Introduction

In the present situation, the volume of generation of municipal solid waste is increasing at very fast rate due to increase in population, industrialization and change in habit and lifestyle of urban population (Hoornweg and Bhada-Tata 1818). Different types of waste—municipal solid waste, electronic waste (E-waste), household waste, etc.—are posing greater challenges to the world. Other than E-waste, the rest generated due to our day-to-day living are considered as municipal solid waste (MSW). In past few decades, researchers and practitioners have shown a significant interest in solving the waste management issues and searched for closed-loop solutions to mitigate the problem. It is evident that better the supply chain of any nation better is the waste management system. Hence, a robust waste management system necessitates a very good reverse logistics network (Baidya et al. 2016). This waste thrown into municipal bins or waste collection centers is collected by the area municipalities for proper disposal. However, either due to resource crunch or inefficient infrastructure and facilities, not all of this waste gets collected and transported to the final dumping sites. If at this stage the management and disposal are not done properly, it can cause serious impacts on health and the problems to the surrounding environment (Ghate and Kurundkar 2016). Municipal solid waste in developing countries like India comprises a big portion of biomass material such as paper, food, wood waste, clothes, plastics, vegetable, rubbers and others daily used discarded materials (Klein 2002). The main functional elements of municipal solid waste management (MSWM) are waste generation, storage, collection, transportation, processing, recycling and disposal in a suitable landfill (Khan and Samadder 2014). Poor collection and inadequate transportation are responsible for the accumulation of MSW at every nook and corner. The management of MSW is going through a critical phase, due to the unavailability of suitable facilities to treat and dispose of the larger amount of MSW generated daily in metropolitan cities. Unscientific disposal causes an adverse impact on all components of the environment and human health (Rathi 2006; Sharholy et al. 2005; Ray et al. 2005). The quantity of MSW generated depends on a number of factors such as food habits, standard of living, degree of commercial activities and seasons. Data on quantity variation and generation are useful in planning for collection and disposal systems (Sharholy et al. 2008). In India, MSW differs greatly with regard to the composition and hazardous nature, when compared to MSW in the Western countries (Gupta et al. 1998). The quantity of MSW generated depends on a number of factors such as food habits, standard of living, degree of commercial activities and seasons. Data on quantity variation and generation are useful in planning for collection and disposal systems (Sharholy et al. 2008). About 90% of waste is currently disposed of by open dumping. Some commonly used methods by which the waste could be managed are as follows: incineration, landfilling and composting. However, these methods are inefficient and harm the environment. There is a great need to move away from the disposal-centric approach and toward the recovery-centric approach of waste management. This paradigm shift requires some level of public participation by regulating and monitoring waste generation and disposal (Narayana

2009). There are many researches on optimization problems such as vehicle routing problem, capacitated vehicle routing problem and vehicle routing problem with time windows have been studied to reduce cost, less emission, serve customers and depot through optimized route (Faccio et al. 2011). However, most of the researches considered static data rather than real-time dynamic data for optimization. To address the above problems, local governments are usually authorized to manage MSW and their laws give them exclusive rights over waste once it has been placed outside. To handle the increasing problems in SWM, ICTs are becoming more significant due to the growing necessities for automated data acquisition, identification, communication, storage and analysis in connection with swift and parallel computing. But, systems that are not using ICTs pose various limitations in terms of site selection, collection monitoring, intelligent recycling, inefficient waste disposal, etc. ICTs can help to overcome these challenges to make a sound SWM system (McLeod et al. 2013). Many systems have been proposed to settle associated issues and maximize waste management efficiency. The system in the literature is classified into three categories: (i) spatial technologies based systems which are founded by using graphical information system (GIS) and/or global position system (GPS) or remote sensing (RS) as main technology. SWM operators adopt these systems to monitor the location of trash bins and collection vehicles during collection (Zamorano et al. 2009). (ii) Identification technologies based systems where barcode or radio frequency identification (RFID) tags are installed with trash bins for tracking identification to determine their location and to acquire the time of collection (Kietzmann 2008). (iii) Data acquisition technologies based systems that contain several sensory elements installed inside trash bins such as image sensor, distance sensor and volumetric sensor to observe its status (Rovetta et al. 2009). And (iv) data communication technologies that are normally used in all previous three kinds of system to facilitate the transmission of captured or analyzed data (Hannan et al. 2015). The present study tries to propose a system in which the system gathers information from the incoming waste at different waste compacting station using image sensing followed by DBMS integrated system takes a decision on the waste final disposal place. The system will be incorporated in the compactor itself. The system will have an algorithm to take the informative decision on the waste. The rest of the paper has been organized into the following sections. Section 2 describes the study methodology in detail. Section 3 discusses and proposes a sustainable solid waste management model using the compactors. The last section concludes the paper.

## 2   Methodology

This study adopts a case study approach. Firstly, a detail literature review was carried out to gauge the waste management practices using ICT and waste compactor uses around the world and the impact of an effective supply chain framework for an effective utilization of waste compactor. Secondly, fifteen field studies were carried out as a case in different location of the city to gauge the present practices and issues in the supply chain framework at both ends of the waste management system. Thirdly, based on the case analysis, waste compactor management system and literature sustainable measures were proposed with integration of DBMS and image sensing methodology.

## 3  Discussion and Analysis

The case studies for the compacting station were carried out in the number of localities falling under Kolkata Municipal Corporation (KMC). Based on the population and waste generation, the number of compactors varied across the city. Fifteen cases were surveyed consisting of 34 compactors. The compactors usually get 40–110 number of handcarts daily. Depending on the area number of handcart increases or decreases, a single handcart can carry 200 kg of mixed waste. The compactor has a capacity of 10 ton; once it gets filled, the compacter is towed to truck and carried for dumping to the landfill site. The below table shows the area which was chosen as a case for analysis with details of number of compactor, number of handcarts received with waste, manpower utilized and number of trips made by each of the compactors for each station (Table 1).

Waste compactors are used to compact the high volume municipal solid waste for an efficient disposal and transportation to the landfill site. All of Kolkata waste collected is disposed of to Dhapa a landfill site situated in eastern fringes of the cities.

**Table 1**  Compactor at various locations

| Sl. No. | Area | No of compactors | Number of handcart received with waste |
|---------|------|------------------|----------------------------------------|
| 1 | Ballygunge | 2 | 100–110 |
| 2 | College Street | 2 | 80–90 |
| 3 | Golf Green | 2 | 80–85 |
| 4 | Kalighat | 4 | 90–95 |
| 5 | Lake Kali Bari | 4 | 85 |
| 6 | Mirza Ghalib Street | 4 | 100–110 |
| 7 | Subodh Mullick Square | 2 | 97–100 |
| 8 | Wireless Park | 2 | 40–50 |
| 9 | Rajdanga | 1 | 50–60 |
| 10 | Southern Avenue | 2 | 90–95 |
| 11 | New Alipore | 2 | 100–105 |
| 12 | Paranshree | 1 | 70–80 |
| 13 | Chetla | 3 | 100–110 |
| 14 | Elgin road | 2 | 90–95 |
| 15 | Belghoria | 1 | 90–95 |

The compactors are being used throughout the world for waste disposal as coined by number of literature. The major practice around the world is to compact the waste in compactors and transfer them to a sanitized landfill site for methane extraction. In developed countries where the waste is segregated at source, the compactor shows the best economical sustainability. The waste in some cases is compacted and transferred to the WTE plant in other countries as, for example, the UK sells its plastic waste to Sweden as a fuel for WTE plant. The segregated waste with high plastic content is compacted and wrapped in plastic film and transported to the designated WTE plant in Sweden. In number of developing countries like India, the compactors are not being used to its full potential making it an unsustainable practice in long run as for the case of Kolkata where the compacted waste is again dumped in a traditional open landfill site, thus wasting the energy which has been utilized for compacting the waste. The compactor station location in Kolkata can be clubbed into three zones, namely market area, park and lake area and residential area. The waste received in each of the areas differs in their composition, though all the waste is mixed waste, but the percentage of organic fragments, plastic, metals, moisture and inert waste varies (Table 2).

The system tries to make the current system more holistic by adapting the present practice of waste collection. The mixed waste collected in handcart will be gauged by utilizing an image sensing system. The waste will be gauged and a predefined

**Table 2** Characteristics of waste

| Location of compactor | Type of waste as input feedstock (Location specific) |
|---|---|
| Market area | Mixed waste:- |
| | (a) Organic: 40–50% |
| | (b) Plastics: 20–30% |
| | (c) Metals: 5–7% |
| | (d) Moisture: High |
| | (e) Others: papers, clay-cup, flesh, glass, vegetable, etc. |
| Park and lake area | Mixed waste:- |
| | (a) Organic: 35–40% |
| | (b) Plastics: 25–30% |
| | (c) Metals: 5–8% |
| | (d) Moisture: medium |
| | (e) Others: leaves, paper cups, glass, small wood logs, bottles, etc. |
| Residential area | Typical mixed waste:- |
| | (a) Organic: 25–30% |
| | (b) Plastics: 20–25% |
| | (c) Metals: 8–10% |
| | (d) Moisture: less |
| | (e) Others: paper cups, clay cups, bottles, batteries, bulbs, etc. |

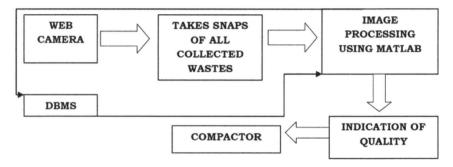

**Fig. 1** Flowchart of the process for decision-making using DBMS-image sensing system (DBMS-ISS)

algorithm will run at background deciding the waste for WTE plant or composting plant. The proposed system utilizes a web camera taking snaps of the waste going into the hopper of compactor. The snap will take a call on the waste characteristics and will decide the compactor in which it should be compacted as from the case study, it was seen that each of the waste compacting stations has two compactors, and this will increase the waste quality after compacting as one of the compactors will have more inorganic fragments and more organic fragments. The image processing technique will utilize MATLAB, running online at background linked with centralized DBMS system incorporating all the waste compactor station thus enhancing a collective decision on the fate of waste (Fig. 1).

The image sensing system takes feedback from DBMS and also gives feedback to the DBMS; it is a two-way feedback system. The system based on the current status of predefined limit of organic and inorganic will take the decision for further processing of the waste toward the WTE plants or composting plants.

The components of collected waste can be identified on the basis of the input as per database. In the database, the properties of any component are taken as its shape, color and surface material. So the waste collected by compactor which consists of different types of organic and inorganic materials can be identified by comparing with any of the features from the existing database, as described below. The feature extraction utilizes the three different methodologies: (i) color feature extraction, (ii) size feature extraction and (iii) texture feature extraction.

(i)   Color Feature Extraction:

In color image, image features in R, G, B color projection values are extracted and stored in database using specified programming methods.

Step 1:- All ten thousand images are taken into working directory of MATLAB.
Step 2:- Using MATLAB programming, all image features in R, G, B color projection values are extracted and stored in database using specified programming methods.
Step 3:- Threshold calculation is taken for categorizing the images into similar feature groups. In this step, threshold value is computed based on the histogram calculation. If the image is a color image, it will convert into gray color image.

Step 4:- In usual methods, image color values are storing in matrix form. Using image matrix, all R, G, B components in image are extracted and separated into three different array forms.

Step 5:- Using the feature vectors, each image color wise means are computed. In this method, row and overall image mean are computed and stored in the database. Based on all these features, different computing methods are formulated.

(ii)   Size Feature Extraction:

Size feature extraction is used to find the size of the vegetable fruit and shape. It depends upon the size parameter of the MATLAB command.

(iii)   Texture Feature Extraction:

These filters are based on multichannel filtering, which emulates some characteristics of the human visual system. The human visual system decomposes an image formed in the retina into several filtered images, each of them having variations in intensity within a limited range of frequencies and orientation.

Thus, the DBMS-ISS will lead to a holistic approach for integrated waste management system for the city of Kolkata addressing the challenges currently the waste management system is having. The system will allow taking an effective decision for the flow of waste to waste-to-energy plant and composting plant, overall making the system more optimized in terms of transportation and inert generation.

# 4   Conclusion

Integrated waste management system is the requirement for a sustainable growth across the city. ICT integration with the waste management system can make the system efficient and more holistic for a city like Kolkata having an infrastructure of exiting compactor station. The DBMS-ISS model has outlined the potential of integrating ICT with waste management system. Further optimization can be done regarding the capturing of image, database upgradation and comparison with collected wasted components, so that the entire process can be fastened. In addition, to extract appropriate results in shortest time, some extra features can be added in the algorithm to make the system more holistic.

# References

Baidya, R., Debnath, B., De, D., & Ghosh, S. K. (2016). Sustainability of modern scientific waste compacting stations in the city of Kolkata. *Procedia Environmental Sciences, 31,* 520–529.

Faccio, M., Persona, A., & Zanin, G. (2011). Waste collection multi objective model with real time traceability data. *Waste Management, 31*(12), 2391–2405.

Ghate, S. S., & Kurundkar, S. V. (2016). SWACHH: An effective real time solid waste management system for municipality. *International Journal of Computer Applications, 149*(4).

Gupta, S., Mohan, K., Prasad, R., Gupta, S., & Kansal, A. (1998). Solid waste management in India: Options and opportunities. *Resources, Conservation and Recycling, 24*(2), 137–154.

Hannan, M. A., Al Mamun, M. A., Hussain, A., Basri, H., & Begum, R. A. (2015). A review on technologies and their usage in solid waste monitoring and management systems: Issues and challenges. *Waste Management, 43,* 509–523.

Hoornweg, D., & Bhada-Tata, P. (1818). *The world bank: What a waste- a global review of solid waste management.* Urban Development & Local Government Unit, World Bank, 1818 H Street, NW, Washington, DC 20433 USA, 2012. Maxwell, J. C (1892). A treatise on electricity and magnetism, 3rd ed., vol. 2. Oxford: Clarendon, pp. 68–73.

Khan, D., & Samadder, S. R. (2014). Municipal solid waste management using Geographical Information System aided methods: A mini review. *Waste Management and Research, 32*(11), 1049–1062.

Kietzmann, J. (2008). Interactive innovation of technology for mobile work. *European Journal of Information Systems, 17*(3), 305–320.

Klein, A. (2002). *Gasification: An alternative process for energy recovery and disposal of municipal solid wastes.* Columbia University.

McLeod, F., Erdogan, G., Cherrett, T., Bektas, T., Davies, N., Speed, C., … & Norgate, S. (2013). Dynamic collection scheduling using remote asset monitoring: Case study in the UK charity sector. *Transportation Research Record: Journal of the Transportation Research Board* (2378), 65–72.

Narayana, T. (2009). Municipal solid waste management in India: From waste disposal to recovery of resources? *Waste Management, 29*(3), 1163–1166.

Rathi, S. (2006). Alternative approaches for better municipal solid waste management in Mumbai, India. *Waste Management, 26*(10), 1192–1200.

Ray, M. R., Roychoudhury, S., Mukherjee, G., Roy, S., & Lahiri, T. (2005). Respiratory and general health impairments of workers employed in a municipal solid waste disposal at an open landfill site in Delhi. *International Journal of Hygiene and Environmental Health, 208*(4), 255–262.

Rovetta, A., Xiumin, F., Vicentini, F., Minghua, Z., Giusti, A., & Qichang, H. (2009). Early detection and evaluation of waste through sensorized containers for a collection monitoring application. *Waste Management, 29*(12), 2939–2949.

Sharholy, M., Ahmad, K., Mahmood, G., & Trivedi, R. C. (2005, December). Analysis of municipal solid waste management systems in Delhi–a review. In *Book of proceedings for the second International Congress of Chemistry and Environment, Indore, India* (pp. 773–777).

Sharholy, M., Ahmad, K., Mahmood, G., & Trivedi, R. C. (2008). Municipal solid waste management in Indian cities–A review. *Waste Management, 28*(2), 459–467.

Zamorano, M., Molero, E., Grindlay, A., Rodríguez, M. L., Hurtado, A., & Calvo, F. J. (2009). A planning scenario for the application of geographical information systems in municipal waste collection: A case of Churriana de la Vega (Granada, Spain). *Resources, Conservation and Recycling, 54*(2), 123–133.

# Motivating Low Carbon Waste Management Through Public–Private Partnerships—An Exploratory Case Study of India

**Tharun Dolla and Boeing Laishram**

**Abstract** Operations of the municipal solid waste (MSW) infrastructure is yet to focus on the low carbon impetus though there is a growing interest of governments and research communities on this topic across the globe. This study addresses and develops propositions concerning Indian infrastructure procurement. More specifically, it draws attention to the advancement of low carbon infrastructure when using public–private partnership (PPP) mode. The findings indicate that low carbon measures can lead to low carbon pathways when the process of executing is favourable to such an agenda. There is clear and outright need to make changes in the way the PPP project is procured to support the low carbon infrastructure (LCI) projects. One such approach is by adopting the criteria for selecting the private concessionaire, appropriating the financial bidding, setting low carbon operational parameters, standards and targets even at the local/project level procurement. This conceptual framework would support the integration of LCI principles and also gives a research trajectory for further research in this area.

**Keywords** India · Low carbon infrastructure (LCI) · Mitigation · Municipal solid waste management (MSWM) · Procurement framework · Public–private partnerships (PPPs)

## 1 Introduction

Sustainable infrastructure provision has grabbed the attention of governments globally. Sustainability is achieved through a two-pronged approach, namely adaptation and mitigation, where without such low carbon pathways development cannot be delivered sustainably (IPCC 2009). For example, Leadership in Energy and Environmental Design, a popular green building certification program is considered to be a sustainability assessment tool (Kamal Mohammad Attia 2013) and also as a climate change mitigation tool (Gunawansa and Kua 2014). So, climate change and

T. Dolla · B. Laishram (✉)
Infrastructure Engineering and Management Division, Department of Civil Engineering, Indian Institute of Technology Guwahati, Guwahati 781039, Assam, India
e-mail: boeing@iitg.ac.in

© Springer Nature Singapore Pte Ltd. 2020
S. K. Ghosh (ed.), *Waste Management as Economic Industry Towards Circular Economy*, https://doi.org/10.1007/978-981-15-1620-7_5_15

sustainable development are inseparable tenants where if at all, a development pathway is progressing without tackling climate change, then it cannot be considered as a sustainable development pathway. Hence, fighting climate change is identified in the UN's Agenda. Many international climate change agreements such as the Paris Agreement encompass voluntary pledges by member states to promote sustainable development by handling climate change. This thinking needs to be infused in the goals and process of any organisation (Wright and Nyberg 2017).

Climate change governance is a continuum of policies that are developed, implemented and encouraged by state and non-state actors (Auld et al. 2014). The next important aspect of approaching this problem is to define and understand the low carbon infrastructure (LCI). Kennedy and Corfee-morlot (2012) describe low carbon infrastructure development as "constructing, or renovating, infrastructure systems (power, road, rail, water, buildings, etc.) to substantially reduce global GHG emissions, while simultaneously making these systems, and the societies they serve, more adaptable to extreme weather conditions and rising sea levels". However, this is a dichotomised definition consisting of mitigation and adaptation. Reduction of carbons is mitigation and making the infrastructure resilient is an adaptation mechanism. Ryan-Collins et al. (2011) argue that mitigation investments will result in cooperative development such that benefits will span beyond just emission reductions. Climate change mitigation activities improve the quality of life of people with measurable benefits (Chomaitong and Perera 2013). In spite of this, climate change mitigation activities did not see much success. Legal agreements like Kyoto protocol have limited enforcement mechanisms. Also, there is evidence that waste and its recycling, water and sewerage are few sectors that saw a decreased trend in global loans to project finance from 2009 to 2011 with $-39\%$ and $-79\%$ decrement, respectively (Kennedy and Corfee-morlot 2012). This trend represents the marginalisation of low carbon projects. Given the extreme paucity of finances to fund the adaptation and mitigation measure of cities, PPPs are believed to be a potential approach for acquiring finances and resources (Dolla and Laishram 2017).

The urban sectors, such as solid waste management, water and wastewater sectors, have also recorded private sector participation in various cities in India. Cities such as Delhi, Bangalore, Coimbatore, Kolkata, Chennai, Ahmedabad, Chennai, Jaipur, Rajkot, Guwahati and Hyderabad have involved private sector through the PPP route in treatment and disposal of waste. Waste to energy projects has been set up in cities such as Hyderabad, Vijayawada, Lucknow, Mumbai and Chennai. The PPP programmes undertaken by the governments at various levels are directed at filling the massive gap between the demand and supply of infrastructure services. However, the infrastructure development programmes have not explicitly integrated the climate change aspect with development goals. Ryan-Collins et al. (2011) have indicated that climate change, though, was perceived as an integral part of the environment, independent attention to it has been given in recent time by the academic community. In India, let alone integrating climate change mitigation, management of MSW had not been given due importance until the intervention by the Supreme Court of India in 1999. This intervention resulted in the formation of an expert committee for streamlining the MSW practices. The recommendations of this expert committee were to enact MSW Rules 2000 for providing the guidelines to various agencies on how to manage MSW. These rules, however, have not given direction and emphasis on adopting measures for climate change mitigation in MSW project design and delivery. The expert report of planning commission (Planning Commission 2014) presents the "co-benefits framework for low carbon strategy" with four categories, namely power, industry, transport and others aspects related to efficiency, codes and forest cover. However, it is silent on the MSW sector.

Solid waste management is amongst the various opportunities in urban development projects to reduce GHGs, making climate change mitigation very relevant to the waste management sector (ADB 2013). Globalisation is resulting in a rapid shift of population to cities, and the many-fold increases in the urban population. Within the anthropogenic activities, cities are the major contributors to these GHGs. Therefore, cities are potentially the best places to achieve GHG mitigation (Anguelovski and Carmin 2011). Copenhagen Climate Change Communiqué notes "…the future of our globe will be won or lost in the cities of the world" (Bulkeley et al. 2011: page 1). In India, around 57.73 million tons of $CO_2$ equivalents ($tCO_2e$) have been estimated to be emitted from this sector, wherein emission from solid waste constitutes about 12.69% (Planning Commission 2014). The emissions from waste have been observed to gradually increase at an annual growth rate (compounded) of 7.3% in the last decade. Assuming the same rate for emissions as well as the GDP, the emissions from the waste sector are estimated to reach from 146 to 163 million $tCO_2e$ in the year 2020 under a scenario of 8% GDP growth rate (Planning Commission 2014). Waste disposal practices are classified into three categories (Kennedy and Corfee-morlot 2012): "low carbon" with less than 0 $tCO_2e$/tonne of waste and includes practices like source separation, recycling and digestion; "medium carbon" has 0–500 kg $CO_2e$/t waste and incineration comes in this category; landfill and open dumping are classified into "high carbon" scenario with greater than 500 kg $CO_2e$/t waste.

PPP is already being used as one of the preferred modes for the development of infrastructure projects relating to highways, hospitals, social housing, airports and dams. However, the extent of application of PPP solely for climate change adaptation and mitigation initiatives, particularly in the urban infrastructure sector, is in the nascent stage. These initiatives are more relevant in the case of urban infrastructure projects such as municipal solid waste (MSW) management projects. In these projects, the focus on project development has been widened in recent times to include climate change mitigation as one of the goals of infrastructure development. Therefore, to manage climate change, research should focus on the identification of the specific challenges and incorporation of various responses to facilitate the management of infrastructure projects (Howard-Grenville et al. 2014). Researchers viewed the development of a mitigation framework for supporting the project performance assessment as an indispensable step. Lind and Borg (2010) argued that traditional business-as-usual behaviour of contractors would drive the projects unless there is a directive or motivation from the public agency. More specifically, the approaches and measures devised by governments dictate the results of PPP projects. But the conventional performance monitoring in these MSW PPP projects is handled by the pollution control board (PCB) who are least equipped with the contractual kind of monitoring. Also, PCB specifies the abstract benchmarks to the MSW projects and is inadequate in operationalising the delivery of low carbon solutions, limiting the power of contracts that empower the PPP model. Such guidance is utterly missing as no agency or government developed directive measures for these projects. This lack points to an urgent requirement for a schema that stimulates PPP processes towards the reduction of carbon emissions. The performance indicators proposed in this study cater to this need. Furthermore, this framework will also be a step towards deliberation for a procurement process as the agenda of climate changes mitigation in its core. This deliberation is achieved by concentrating on the selection criteria of private entities until the operational obligations spelt in contracts. The objective of this study, in the broad context of sustainable development, is a framework for PPPs using LCI principles as a theoretical lens to support the Indian MSW PPP projects.

## 2   Conceptual Background

Public procurement accounts for more than a fourth of India's GDP and amounts to ₹28 trillion, having a significant influence on the economy's performance. Public procurement in India is a framework with constitutional provisions (articles of the constitution), legislative provisions (acts of parliament), administrative guidelines (rules and manuals) and overseers (e.g. Central Vigilance Commission) as the building blocks. It is viewed as a more sophisticated form of governance. Public procurement holds the key to impact the policy to attain a low carbon society with the help of circular economy principles. As the responsibility of public infrastructure provision is with the governments, they have adopted various modes of project delivery like partnering, and PPP/PFI to involve the private sector in the creation of infrastructure assets and provision of infrastructure services. Currently, developing countries take the lead in using and researching PPP mode, and this leaning is expected to continue for another 20 years (Akintoye et al. 2016). Municipal infrastructure deficit is more increasingly filled by PPP mode (Akintoye and Kumaraswamy 2016). Though some success stories of PPP projects have been driving PPPs, care should be warranted in ascribing success to PPP mode of procurement. For instance, the private investment, which is the core of PPP, is contested on its contribution to sustainability (Koppenjan 2015). This contestation shows the danger of PPPs and also raises the attention to the PPP route of infrastructure development across the sectors, so much that sustainable development goals are achieved. Moreover, small-scale public utility PPP projects such as municipal solid waste projects in developing economies draw more urgency due to the lack of research studies.

Usage of PPPs in the MSW sector has added come more challenges to the set. The incomplete nature of the PPP contracts provides leeway to the occurrence of contractual hazards. As a result, execution and the scrutiny of the projects become more tedious. One way of solving this problem is by letting the stakeholders expand contractual obligations by relating to performance (Klijn and Koppenjan 2016). This expansion can be achieved by knotting penalties with the contract clauses to stimulate better enforcement in PPP projects. Thus, a monitoring mechanism that is dependent on and informed by a contractual framework and having performance indicators as the building blocks (Williamson 1999) provides the way forward for this performance measurement objective. Consequently, developing an appropriate mechanism with emphasis on criteria of PPP performance by the help of quantifiable indicators holds the key (Akintoye and Kumaraswamy 2016), thereby satisfying the public–private stakeholder groups of PPP, including the end-users. In this context, social, political, environmental and economical are driving the implementation success of PPP projects and demands meticulous deliberation. PPP MSW projects have a nature of complexity on a relative scale when compared to other infrastructure sectors on account of the multitude of technologies available for treatment with varying costs and the need for active participation of people belonging to different social strata. Also, MSW projects have a broader implication on the society and are socially non-excludable (unlike roads, railroads, power and other sectors which are socially excludable) leading to widespread political consequence. This sector also has an extreme environmental challenge. Thus, the key stakeholders, such as public agencies and private entities of PPP MSW projects, would require to address these issues and the multifaceted interactions by using most appropriate measures.

LCI philosophies can be incorporated in the procurement regulatory framework of PPPs using performance indicators (Patil et al. 2015). The basis for tariff changes and the modalities of regulations are some of the issues of PPP governance in connection with the monitoring of performance (Delhi and Mahalingam 2013). As a result, there raised a need in the studies related to PPP to focus more acutely on the performance indicators (Akintoye and Kumaraswamy 2016). Such support is conceivable by developing ex-ante evaluation tools having the capability to increase confidence and transparency while procuring these projects, especially in PPP mode (Akintoye and Kumaraswamy 2016). Likewise, there requires a policy framework by governments on climate change that enables unified planning for infrastructure sectors to back up the policy decision that customarily establishes the mitigation objectives in the process of fight climate change (Ryan-Collins et al. 2011). Such an action plan-based regulatory framework that is grounded on LCI principles will redirect the conventional economic style of decision-making. In this process, the tensions that arise between the modalities of PPP mode and the newly added LCI agenda would be addressed (Koppenjan 2015). Past studies were limited to the compliances, payments and value for money aspects of PPPs and are inadequate to the performance issues in the PPP in the light of LCI and sustainable development agenda—making this focus very significant. Agility and ability to account the unanticipated problems related to climate change become essential in the benchmarking of PPP performance along with the contractual monitoring of the project during the concession period. Constructing performance indicators of PPP, which can encourage a low carbon solution during the delivery by the private sector, becomes essential in the procurement process.

# 3   Methodology

The research question of this study was to know the means to bolster the inclusion of LCI principles in the PPP procurement process. This research falls into the domain of "How" kind of inquiry and makes the case study methodology most apt. Six key dimensions—selection criteria, selection process, financial mechanism, operational aspects, standards and targets—form the analytical framework derived from the extant literature. A PPP MSW project of India is selected as a test case. India stood as the largest PPP market in terms of private sector participation in infrastructure amongst the developing countries in the period 2008–2012. Thus, a case study from such an emerging market would provide a representative PPP market behaviour and therefore is of much relevance to other developing economies. Semi-structured interviews with five stakeholders provided the primary evidence for the case study while the project-related documents provided the secondary evidence for corroborative purposes. The five interviewees' had 19.6 years of average experience, indicating the richness of the data. After analysing the presence of LCI principles in the case study through primary and secondary data, the evidence collected by using the analytical framework was further validated through the second round of interviews with the earlier PPP MSW stakeholders.

# 4   Analysis and Discussion

To analyse the level of LCI principles in the project, practices of the case study organisation (s) are scrutinised. The summary of evidence from this case study project concerning the dimensions of the analytical framework is presented in Table 1.

# 5   Suggestions and Way Forward

In the conventional mode of procuring the PPP projects, ULB looks for cost-effective financial proposals jeopardising the importance of soundness of technicalities for emission reductions. This evaluation requires a weightage to technological soundness in the selection process as assessed from the climate-friendly perspective and the impact of the environment. One expert states, "…we have to understand that quality treatment is must and economics should not be the first criteria of selection of technology nor a PPP project concessionaire". Some measures suggested to achieve this objective are: there is a need that ULBs should attempt understanding the mitigation potential of the proposed technologies. This understanding requires robust quantification studies. Additionally, these technologies can be weighed based on output specifications when selecting the concessionaire. Since the payment mechanism is

**Table 1** Summary of the case study evidence

| Dimension | Low carbon PPP MSW project requirements | Case study evidence |
|---|---|---|
| Selection criteria | Carbon metrics: Incorporation of low carbon-related parameters in the criteria of evaluation (Schaltegger and Csutora 2012; Jowitt et al. 2012) | Carbon emissions are not accounted for in the decision-making process. The reasons are attributed to lack of knowledge and expertise for such actions |
| | Performance-based output specifications (PBOS): LCI requires not just a financially sustainable bid in final bidding criteria but a whole set of criteria (Kennedy and Corfee-morlot 2012; Koppenjan 2015). Criteria supporting innovation should be incorporated. Innovation may not be with technology change alone where hi-tech developments are a subset of low carbon innovation (Tyfield and Jin 2010). It requires freedom to the private sector to recommend their preferred technology based on studies such as waste characteristics conducted for bidding and to comply with PBOS | The project is awarded only based on the tipping fee. The environmental requirements are considered for meeting the requirements only but not in the selection of bid. Criteria are not supporting innovation and innovative solutions. Currently, the technology is recommended at the level of DPR in the project, and the private sector builds on the DPR in their proposals |
| Selection process | Technology innovation: Low carbon innovative technology is a necessity for mitigating emissions in LCI (Koppenjan 2015) | No means to promote innovative technology and enhance competition in the provision of low carbon solutions |
| | Stakeholders involvement: Incorporating stakeholder involvement and ownership aiming at coordination of contracting parties and stakeholders interests alongside with LCI objectives (Koppenjan 2015; Kuronen et al. 2010) | Lack of stakeholder analysis and management caused PPP contract to fail and subsequently the poor private investment in LCI by the private sector. Decisions made by ULB and transaction advisory in the early stages of procurement are passed down to the private sector for implementation representing a top-down approach |

(continued)

**Table 1** (continued)

| Dimension | Low carbon PPP MSW project requirements | Case study evidence |
|---|---|---|
| Financial mechanism | Performance-based payments: Payments concerning the emissions reduced in the period of accounting (Schaltegger and Csutora 2012; Jowitt et al. 2012). The private sector would build facilities that can divert waste from landfill with innovative thinking if landfill tax is implemented. | Fixed payments to the private sector as agreed in an accounting period. Their exists no motivation for innovation to achieve performance-related payments |
| | In cases where a positive business case is absent. Government has to provide a guarantee and redesigning the project in the way of value capturing (other sources of revenue) and scope management (development rights, enhancing the scope of business of operator, etc.) (Koppenjan 2015). Capital subsidies and loan schemes encourage private sector for investing in low carbon projects (Wong et al. 2013) | The current PPP projects are mostly funded by the government by giving an upfront grant for the project without any consideration of technology innovation |
| Operational aspects | Providing flexibility to continually upgrade the technology (Guasch 2004) | No flexibility in technology and innovation during the operation phase within the concession period |
| Standards | Promotion of recycling and markets for recycling. Segregation at source will help recycling as well as treatment of waste with the most suitable low carbon option (Kennedy and Corfee-morlot 2012) | No linkages for the promotion of recycling and no emphasis on low carbon options |
| Targets | Setting up emission reduction targets in the MSW sector drives the projects in the direction of reaching the goals (Corfee-Morlot and Höhne 2003) | Policy-based reduction of emission is missing at the local level |

so critical in PPPs, reorienting by suitably relating to emission reductions will facilitate and motivate competencies for the LC projects. Low carbon instruments such as landfill tax on the private concessionaire can act as motivation for the private sector. Nevertheless, though practised in the UK, implementation of this tool in India is not without resistance in the present state of affairs. Formulation of regulations, policies, codes and support programs are the core of instituting urban climate activities

(Anguelovski and Carmin 2011). Expert committees that are formed in the purview of waste management did spur significant progress in all dimensions of MSW in India. Still, pursuing the problem with policies and mechanisms in climate change mitigation perspective appears to be absent.

The funding deficit, when provided as an upfront grant, failed to motivate the private sector to operational responsibilities and provided leeway for contractual failure due to the operational risk aversion. Instead, when outputs of the project are linked with the operational grants for the deficit funding or situations of higher operational costs, there is an increase in the effectiveness of the PPP contracts. Renegotiation clauses are often contested by Indian governments in the normal circumstances, but LCI project requires such flexibility to adopt new technological insights and developments. To avoid the discrepancies, probity arrangements such as ratification of the technological advancement will reinforce the motivations. As transaction costs are expected to increase, there should be limited flexibility in renegotiation clauses.

Waste reduction and promotion of efficiency are enhanced by the decentralised manner of dry and wet waste collection and treatment. The current design of PPP is deterrent to such practices. Subsequently, the cost of treatment increases as the waste goes farther from the point of generation. It also requires more extensive mechanical pre-treatment facilities. Some practical ways to reduce carbon emissions are to separate waste at collection points, segregation of dry recyclables, composting the organic waste, substitute fertiliser with compost, controlled landfilling of inert (Couth and Trois 2010). PPPs need to reconsider these aspects to make the projects more of low carbon nature. There is a need to create a task force comprising public and private entities that can spur climate change mitigation in urban infrastructure sectors such as energy, water supply, waste management, natural resource conservation, transportation and communications infrastructure (Rosenzweig et al. 2010). The responsibility of ULBs for emission reductions is believed to be enhanced when the central government sets up suitable targets for the state and local bodies in the aspects such as waste reduced, diverted from landfill, localised waste treatment facilities. For this, there is a need to set up a systematic review at the zonal level for the sake of projects. The waste sector needs to focus on more sustainable approaches and resource recovery by implementing the principles of the circular economy.

# 6   Conclusions

Municipal waste management sector is one of the potentials but neglected sectors in the low carbon infrastructure ambition. Having provided with PPP mode of procurement, this public procurement projects need to be highly motivated to create a conducive environment for emission reductions through the supply chain operations of MSW. This context requires modification of procurement methodology in the aspects such as adapting the criteria for selecting the private concessionaire, appropriating the financial bidding, setting low carbon operational parameters, standards and targets even at the local/project level procurement. This study recommends strategies such as capability upscaling of the public bodies to help understand and implement the climate change mitigation processes. Essential tools in this regard are the quantification of the potential of technologies for emission reductions, to develop more informed choices. Viability gap funding failed to meet its purposes and hence should be encouraged in the operations rather than in construction. The central government and state government need to set emission reduction targets and promote sustainability reporting. The preliminary contributions of this study require further validation through multiple cases and develop the best practice framework of low carbon infrastructure procurement through PPPs.

**Acknowledgements** The authors acknowledge the funds received by Housing and Urban Development Corporation's (HUDCO) Human Settlement Management Institute, New Delhi, India, for carrying the research titled "A Study on Feasibility of PPPs as Mitigation Strategy for Climate Change". This publication and all other working papers/journals/conferences papers form the part of that project. The authors wish to express their sincere gratitude to all practitioners participated in this research study.

# References

ADB. (2013) *Financing Low-Carbon Urban Development in South Asia: A Post-2012 Context.* Asian Development Bank.

Akintoye, A., & Kumaraswamy, M. M. (2016). *Public-Private Partnerships CIB TG72 Research Roadmap.* CIB General Secretariat.

Akintoye, A., Beck, M., & Kumaraswamy, M. M. (2016). *Public-Private Partnership: A Global Review.* Routledge.

Anguelovski, I., & Carmin, J. (2011). Something borrowed, everything new: Innovation and institutionalization in urban climate governance. *Current Opinion in Environmental Sustainability, 3,* 169–175. https://www.doi.org/10.1016/j.cosust.2010.12.017.

Auld, G., Mallett, A., Burlica, B., Nolan-Poupart, F., & Slater, R. (2014). Evaluating the effects of policy innovations: Lessons from a systematic review of policies promoting low-carbon technology. *Global Environmental Change.* https://doi.org/10.1016/j.gloenvcha.2014.03.002.

Bulkeley, H., Broto, V. C., Hodson, M., & Marvin, S. (2011). *Cities and Low Carbon Transitions.* Routledge. https://doi.org/10.1080/08111146.2013.845133.

Chomaitong, S., & Perera, R. (2013). Adoption of the low carbon society policy in locally-governed urban areas: experience from Thai municipalities. *Mitigation and Adaptation Strategies for Global Change,* 1255–1275. https://doi.org/10.1007/s11027-013-9472-0.

Corfee-Morlot, J., & Höhne, N. (2003). Climate change: long-term targets and short-term commitments. *Global Environmental Change, 13,* 277–293. https://www.doi.org/10.1016/j.gloenvcha.2003.09.001.

Couth, R., & Trois, C. (2010). Carbon emissions reduction strategies in Africa from improved waste management: A review. *Waste Management, 30,* 2336–2346. https://www.doi.org/10.1016/j.wasman.2010.04.013.

Delhi, V. S. K., & Mahalingam, A. (2013) A framework for post award project governance of public-private partnerships in infrastructure projects. In P. Carrillo & P. Chinowsky (Eds.), *Engineering Project Organization Conference Woking Paper Proceedings.* Colorado.

Dolla, T., & Laishram, B. (2017). Public-private partnerships for climate change mitigation—An Indian case. *MATEC Web Conference, 120,* 02022. https://www.doi.org/10.1051/matecconf/201712002022.

Guasch, J. L. (2004). *Granting and Renegotiating Infrastructure Concessions: Doing it Right.* The World Bank. https://doi.org/10.1596/0-8213-5792-1.

Gunawansa, A., & Kua, H. W. (2014). A comparison of climate change mitigation and adaptation strategies for the construction industries of three coastal territories. *Sustainable Development, 22,* 52–62. https://www.doi.org/10.1002/sd.527.

Howard-Grenville, J., Buckle, S. J., Hoskins, B. J., & George, G. (2014). Climate change and management. *Academy of Management Journal, 57,* 615–623. https://www.doi.org/10.5465/amj.2014.4003.

IPCC. (2009). *Concept paper for an IPCC Expert Meeting on Human Settlement, Water, Energy and Transport Infrastructure—Mitigation and Adaptation Strategies.* Intergovernmental Panel on Climate Change.

Jowitt, P., Johnson, A., Moir, S., & Grenfell, R. (2012). A protocol for carbon emissions accounting in infrastructure decisions. *Proceedings of the Institution of Civil Engineers: Civil Engineering, 165,* 89–95 (2012). https://www.doi.org/10.1680/cien.2012.165.2.89.

Kamal Mohammad Attia M (2013) LEED as a tool for enhancing affordable housing sustainability in Saudi Arabia:The case of Al-Ghala project. Smart and Sustainable Built Environment 2(3): 224–250. https://www.doi.org/10.1108/SASBE-02-2013-0009.

Kennedy, C., & Corfee-morlot, J. (2012). *Mobilising Investment in Low Carbon, Climate Resilient Infrastructure. OECD Environmental Working Paper.* The Organisation for Economic Co-operation and Development (OECD) Publishing.

Klijn, E. H., & Koppenjan, J. (2016). The impact of contract characteristics on the performance of public-private partnerships (PPPs). *Public Money & Management, 36,* 455–462. https://www.doi.org/10.1080/09540962.2016.1206756.

Koppenjan, J. F. M. (2015). Public–Private Partnerships for green infrastructures. Tensions and challenges. *Current Opinion in Environmental Sustainability, 12,* 30–34. https://www.doi.org/10.1016/j.cosust.2014.08.010.

Kuronen, M., Junnila, S., Majamaa, W., & Niiranen, I. (2010). Public-private-people partnership as a way to reduce carbon dioxide emissions from residential development. *International Journal of Strategic Property Management, 14,* 200–216. https://www.doi.org/10.3846/ijspm.2010.15.

Lind, H., & Borg, L. (2010). Service-led construction: Is it really the future? *Construction Management and Economics, 28,* 1145–1153. https://www.doi.org/10.1080/01446193.2010.529452.

Patil, N. A., Tharun, D., & Laishram, B. (2015) Infrastructure development through PPPs in India: Criteria for sustainability assessment. *Journal of Environmental Planning and Management, 59*(4), 708–729. https://www.doi.org/10.1080/09640568.2015.1038337

Planning Commission. (2014). *The Final Report of the Expert Group on Low Carbon Strategies for Inclusive Growth.* Government of India.

Rosenzweig, C., Solecki, W., Hammer, S. A., & Mehrotra, S. (2010). Cities lead the way in climate-change action. *Nature, 467,* 909–911 (2010). https://www.doi.org/10.1038/467909a.

Ryan-Collins, L., Ellis, K., & Lemma, A. (2011). *Climate Compatible Development in the Infrastructure Sector.* Institution of Civil Engineers (ICE).

Schaltegger, S., & Csutora, M. (2012). Carbon accounting for sustainability and management. Status quo and challenges. *Journal of Cleaner Production, 36,* 1–16. https://www.doi.org/10.1016/j.jclepro.2012.06.024.

Tyfield, D., & Jin, J. (2010). Low-carbon innovation in China – introduction to the special issue. *Journal of Knowledge-based Innovation in China, 2.* https://www.doi.org/10.1108/jkic.2010.40402caa.001.

Williamson, O. E. (1999). Public and private bureaucracies: A transaction cost economics perspectives. *The Journal of Law, Economics, and Organization, 15,* 306–342. https://www.doi.org/10.1093/jleo/15.1.306.

Wong, P. S. P., Ng, S. T. T., & Shahidi, M. (2013). Towards understanding the contractor's response to carbon reduction policies in the construction projects. *International Journal of Project Management, 31,* 1042–1056. https://www.doi.org/10.1016/j.ijproman.2012.11.004.

Wright, C., & Nyberg, D. (2017). An inconvenient truth: How organizations translate climate change into business as usual. *Academy of Management Journal, 60,* 1633–1661. DOI: 10.5465/amj.2015.0718.

# Opportunities for Resource Recovery After Hydrothermal Pretreatment of Biodegradable Municipal Solid Waste: A Mini-review

**Divya Gupta, Sanjay M. Mahajani and Anurag Garg**

**Abstract** The disposal of biodegradable fraction of municipal solid waste (MSW) is a major problem worldwide. Composting and anaerobic digestion are suggested as potential treatment methods for such a waste. Due to long processing time, odor emissions and susceptibility to toxic metals, these processes have received limited success at majority of places. Therefore, there is enhanced interest in hydrothermal carbonization (HTC) process for the pretreatment of biomass waste. HTC is conducted at 180-260°C in the presence of moisture and auto-generated pressure for few minutes to hours to form a carbon-rich solid mass (also known as hydrochar) and wastewater with high organics concentration. Hydrochar has high heating value (>20 MJ/kg) and can be used as co-fuel. Alternatively, it can be utilized as soil conditioner. The wastewater can be used for biogas recovery, carbohydrate recovery or bioethanol formation via fermentation. Around 90% carbon recovery is possible in solid and liquid fractions after HTC pretreatment. Establishment of decentralized-scale plants for high moisture wastes (>70%) such as, kitchen waste, garden trimmings, sewage sludge, institutional wet waste, and food processing industrial waste can help in recovery of energy from waste at source. However, optimization of the reaction conditions for different wastes and analysis of wastewater characteristics to assess its recovery potential are yet to be performed. In this mini-review, the recent literature on HTC of MSW, possible reactions during HTC and final products are discussed and gaps in the existing information are highlighted.

**Keywords** Biodegradable solid waste · Hydrothermal carbonization · Resource recovery · Waste to energy · Municipal solid waste management

D. Gupta · A. Garg (✉)
Environmental Science and Engineering Department, Indian Institute of Technology Bombay, Mumbai, India
e-mail: a.garg@iitb.ac.in

S. M. Mahajani
Department of Chemical Engineering, Indian Institute of Technology Bombay, Mumbai, India

© Springer Nature Singapore Pte Ltd. 2020
S. K. Ghosh (ed.), *Waste Management as Economic Industry Towards Circular Economy*,
https://10.1007/978-981-15-1620-7_5_16

149

# 1  Introduction

Disposal of wet biodegradable organic fraction of municipal solid waste (OFMSW) is a major challenge in the developing world. Composting and anaerobic digestion (AD) are considered the two major biological options for resource recovery in the form of fertilizer-type material and biogas, respectively. However, there are three major issues: high processing time, biological process inhibition due to the presence of toxic compounds (Stemann et al. 2013), and odorous air emissions. Moreover, municipal solid waste (MSW) is a heterogeneous mass having varied properties like moisture content, irregular physical shape, chemical composition, and heating value. To overcome these limitations, pretreatment of MSW can be carried out before it can be utilized as an efficient energy resource (Kambo and Dutta 2015).

Hydrothermal treatment (HT) of wet waste can be a potential pretreatment alternative to reduce the processing time and enhance resource yield. In recent years, the process is receiving considerable attention worldwide (Berge et al. 2011; Parshetti et al. 2014). The wet biodegradable fraction, mainly kitchen waste (cooked as well as uncooked) and garden trimmings, constitutes a large portion of MSW (38–72%) in major cities of India (Garg 2014). This kind of feedstock is suitable for hydrothermal pretreatment. According to Basso et al. (2016), HT is classified into three categories: carbonization (180–250 °C), liquefaction (250–373 °C), and gasification (>373.95 °C and >22.06 MPa pressure). A classification of HT processes is illustrated in Fig. 1.

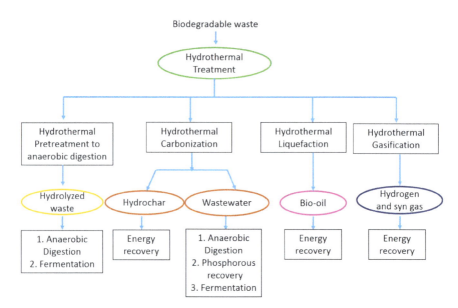

**Fig. 1** Various HT processes for treatment of biodegradable waste and expected end-products

Among these processes, hydrothermal carbonization (HTC) is carried out at least severe reaction conditions and produces solid hydrochar as the main product, which can assist in recovering most of the energy from waste material with lesser input.

The present mini-review paper discusses HTC pretreatment of OFMSW, reactions likely to occur during the process and principle outputs. To accomplish the objective of present paper, review of the studies carried out in last ten years was performed. Finally, the future research needs are highlighted in this area.

# 2 HTC Process

## 2.1 Definition and End-Products

HTC takes place at medium temperature conditions (180–260 °C) under self-pressure (Berge et al. 2011; Lu et al. 2011; Prawisudha et al. 2012; Kim et al. 2012). After the process, gas or liquid products (e.g., biogas or ethanol) can be obtained from wastewater after suitable post-treatment methods, while solid (hydrochar) can be utilized as soil conditioner or co-fuel (Libra et al. 2011). The hydrochar has high calorific value (similar to brown coal), while the wastewater can be subjected to anaerobic biological process or fermentation to obtain biogas or ethanol. If hydrochar is used as fuel, $CO_2$ emitted from its combustion would not contribute towards greenhouse gas (GHG) emissions thus reducing carbon footprint. Char added to soil may also improve crop productivity. Glaser et al. (2001) reported the enhancement in the nutrient holding capacity of soil due to black carbon which increased its productivity.

HTC has been used recently for different biomass materials (mainly ligno-cellulosic), for example, sewage sludge, wheat straw, grape marc, and corn silage. However, only few studies are available on MSW which are also published in last ten years (Yoshikawa 2009; Berge et al. 2011; Lu et al. 2011, 2012; Hwang et al. 2012; Li et al. 2013; Reza et al. 2016). HTC reactions are exothermic and the process can be considered self-sustaining to some extent. Zhao et al. (2014) have reported that it only needs 40% of the energy required for drying the feedstock prior to other thermal processes.

HTC can be applied for heterogeneous feedstock and convert it to a homogeneous powdered product. Being a thermal process, it is not affected by variability in feed. Moreover, it requires lesser time and space as compared to the conventional biological processes. HTC has potential to convert highly moisturized waste OFMSW into carbon-rich product (Kambo and Dutta 2015). Therefore, the process can be used at decentralized facilities, such as institutions, restaurants, and food processing industries.

## 2.2   Reactions Occurring During HTC

The process takes place under auto-generated pressure which ensures that water remains in the liquid form. Under subcritical conditions, water has an increased ionic product and acts as an organic solvent (Kambo and Dutta 2015). Temperature and reaction time mainly govern the solid yield and its calorific value. In previous studies, reaction time has been widely varied from 30 min to 5 days (Lu et al. 2011, 2012; Prawisudha et al. 2012). Therefore, optimization of the reaction parameters needs to be done for different feedstock.

Various reactions likely to occur during HTC are shown in Fig. 2, although the rate of their occurrence is not well known (Kambo and Dutta 2015). Biomass is mainly comprised of three biological compounds: hemicellulose, cellulose, and lignin. Hemicellulose and cellulose hydrolyze at ~180 °C and ~200 °C, respectively (Funke and Ziegler 2010). Their hydrolysis leads to the formation of carbohydrates, volatile fatty acids (VFAs), and 2,5-hydroxymethylfurfural (HMF). HMF gets deposited into the hydrochar structure and increases its energy density which is higher in comparison to the biochar formed during pyrolysis. Subcritical water lowers the activation energy for hemicellulose and cellulose hydrolytic reactions (Kambo and

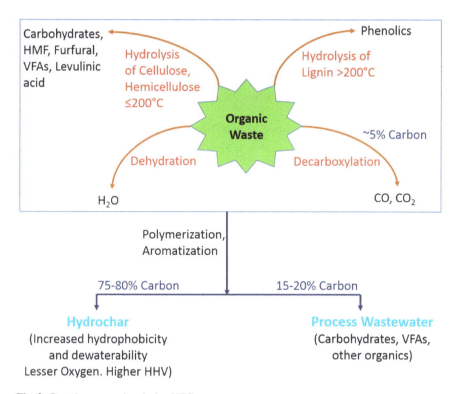

**Fig. 2**  Reactions occurring during HTC

Dutta 2015). Due to lower viscosity of water, destruction of colloidal structures and less hydrophilic functional groups is more likely with gas formation ($CO_2$ and CO). Funke and Ziegler (2010) suggested that initial rate of reaction was higher in alkaline conditions, whereas further degradation rate was enhanced in acidic medium. Oxygen content in the feedstock is reduced by dehydration and decarboxylation reactions (Kambo and Dutta 2015). Low molecular weight compounds are formed which precipitate via condensation polymerization reactions and form hydrochar. Dehydration reaction increases the hydrophobicity and dewaterability of the product. After dehydration, decarboxylation occurs along with removal of phenolic groups from lignin particularly above 200°C temperature. Subsequently, polymerization of unsaturated compounds formed after decarboxylation occurrs (Funke and Ziegler 2010). Aromatization also takes place simultaneously and increases with increasing reaction severity.

Most of the carbon input (75–80%) is reserved in the solid phase (hydrochar), about 15–20% is dissolved in the liquid phase, and the remaining 5% is converted to gas (mainly $CO_2$) (Berge et al. 2011; Hwang et al. 2012). In case of dry thermal treatments, inorganic components such as alkali and alkaline earth metals form clinker, ash, etc., that cause fouling, slagging or corrosion while burning the biochar. However, acetic acid formed during the HT solubilizes these components into the aqueous phase and reduces the ash content in the hydrochar (Funke and Ziegler 2010).

## 2.3  Studies on HTC of MSW

The studies on HTC of MSW vary widely in terms of reaction conditions and focus mainly on the solid product characteristics (Lu et al. 2011; Kim et al. 2012; Parshetti et al. 2014). The initial moisture content of waste ranges from 50–80% and reaction temperature from 160 to 295 °C (Table 1). The studies using lower reaction temperature (<180 °C) aim at hydrolysis of the solids as a pretreatment to AD process. After HTC, increase in calorific value, carbon percentage, and fixed carbon is reported in all the studies with removal of volatile matter (Berge et al. 2011; Lu et al. 2011; Hwang et al. 2012). Hwang et al. (2012) observed that plastics do not degrade during HTC and may lead to overestimation of the final carbon yield. In addition to this, it was observed that organic chlorine is converted to inorganic chlorine during the process, which can be washed off with distilled water. Therefore, problems from the release of dioxins and furans (chlorinated organics) due to burning of feedstock are eliminated. Prawisudha et al. (2012) reported decrease in chlorine content from 10,000 ppm to 2000 ppm. Burning hydrochar can help in extracting more energy than AD, landfill site, and incineration of food waste, while lesser emissions are expected from the process (Lu et al. 2012). Analysis of gas generated during HTC of MSW exhibited the release of furan upto 30 h of reaction duration, after which it was negligible and propene emerged as the most abundant gas (Lu et al. 2012). Amongst paper, food waste, and mixed MSW, food waste-derived hydrochar had the maximum high heating value (30 MJ/kg after 120 h duration), whereas mixed MSW had

**Table 1** Summary of recent studies carried out on HTC of MSW

| Feedstock | Process conditions | Key findings | References |
|---|---|---|---|
| Simulated mixed MSW | 250 °C, 20 h | • 49–75% carbon recovery in char<br>• High heating value increased to upto 29 MJ/kg | Berge et al. (2011) |
| Simulated Japanese, Chinese and Indian mixed MSW | 220 °C, 30 min | • Upto 1.4 times increase in high heating value and 6.4–9 times increase in energy per unit volume<br>• Maximum volatile matter loss (41.8%) in food waste with increase in fixed carbon<br>• Combustion is dominated by component present in major quantity | Lu et al. (2011) |
| Mixed MSW | 234 and 295 °C, 150 min | • 75% carbon recovery in char. Plastics do not degrade during HTC causing overestimation of yield and carbon recovery<br>• Organic chlorine was converted to inorganic chlorine which could be washed easily | Hwang et al. (2012) |
| Mixed MSW | 215–235 °C, 30–90 min | • Steam was supplied to maintain temperature and pressure in reactor<br>• Chlorine reduced from 10,000 (initial) to 2000 ppm in washed hydrochar. Removal increased with increasing reaction severity. No chlorine was found in process wastewater<br>• Maximum high heating value of 24 MJ/kg was obtained<br>• Energy balance showed that 25% of energy contained in hydrochar was required for HTC | Prawisudha et al. (2012) |

(continued)

**Table 1** (continued)

| Feedstock | Process conditions | Key findings | References |
|---|---|---|---|
| Biodegradable waste residue from mechanical segregation | 180–220 °C, 30 min | • 40 and 60% cellulose decomposed at 200 and 220 °C. ~90 and 99% hemicelluloses were decomposed at 180 and 220 °C. Lignin did not degrade significantly<br>• Dehydration and natural drying performances were significantly improved<br>• Increase in calorific value from 14.7 MJ/kg to 21.7 MJ/kg at 200 °C | Kim et al. (2012) |
| Food waste | 250 and 350 °C, 20 min | • Adsorption of two textile dyes using hydrochar was studied which fitted well with Langmuir isotherm and pseudo second-order model<br>• Higher dye removal (95%) obtained at lesser reaction severity (250 °C)<br>• pH and temperature of adsorption experiments significantly affected dye removal | Parshetti et al. (2014) |
| Food waste | 225–275 °C, 4–96 h | • Proper segregation of biodegradable wastes and management of liquid streams will reduce environmental impact<br>• Wastewater disposed into surface waters may cause cancerous and non-cancerous toxicity due to metals as well as eutrophication due to presence of nutrients<br>• Combustion of hydrochar is highly beneficial in reducing global warming potential and acidic potential as compared to coal | Berge et al. (2015) |
| Mixed MSW | 210–280 °C, 30–90 min | • Compared combustion, pyrolysis and gasification of hydrochar<br>• High reaction severity led to increase in ignition temperature in hydrochars due to loss of volatile matter and increase in fixed carbon. Reduced gasification time and increased activation energy of MSW | Lin et al. (2016) |

the minimum. Li et al. (2013) found that increase in packaging material along with food waste decreased the energy recovery due to its low energy retention property. Reza et al. (2016) treated autoclaved OFMSW and the digestate produced after AD of OFMSW by HTC process at 200–300 °C temperature. High heating value and solid yield need to be maximized by optimizing the reaction conditions. Increase in percent concentration of extractives, hemicellulose, and lignin was found with decrease in cellulose content after HTC treatment. Calcium was found in maximum concentration of ~35 to 40 mg/kg of solids amongst various inorganic elements such as Na, Ca, Mg, Al, S, P, Fe, and K. Wastewater after HTC of OFMSW contained organic acids such as propionic, oxalic and acetic, whereas wastewater from HTC of anaerobic digestate resulted in the formation of organic acids as well as amino acids such as L-proline and glycine (Reza et al. 2016).

## 2.4 Gaps Identified in the Research Area

It is mentioned previously that the mechanism of HTC is quite complex and is not well understood. The future work may look upon this aspect and efforts should be made to establish HTC reaction mechanism for OFMSW. Optimized reaction conditions should be determined for OFMSW and sorted MSW as variation in MSW composition affects the overall performance of HTC process. Reaction kinetic models need to be developed considering various reactions occurring during HTC. For hydrochar obtained from OFMSW, the work should be extended to evaluate their performance in pilot-scale systems as mono- and co-fuel. The potential of hydrochar as soil conditioner at pilot level needs to be assessed.

Almost all the studies analyzed hydrochar properties and energy outputs from the process. However, utilization of wastewater has not been well reported. The wastewater has high organic load and acidic pH; therefore, its suitability for fermentation or anaerobic digestion process should be ascertained. Depending upon the feedstock, the liquid waste may have several compounds such as organic acids, carbohydrates, phosphorous, ammonium, phenols, etc. Major compounds generated during HTC of biomass have been identified in wastewater in several studies (Becker et al. 2014; Kambo and Dutta 2015). Similar analysis of wastewater from HTC of OFMSW should be performed to assess its usage and possible resource recovery routes. Pretreatment with dilute acids or physicochemical processes can increase sugar yield and bioethanol formation (Mohapatra et al. 2017). AD is suggested to be economically feasible for HTC derived wastewater at an industrial scale if chemical oxygen demand (COD) degradation is more than 80% and organic loading rate is greater than 1.5 kg $COD/m^3/d$ (Wirth et al. 2012). It can also be used as a liquid fertilizer as it contains nutrients such as N, P, and K and heavy metals are concentrated in the solid residue (Yoshikawa 2009). The concern for heavy metals may be reduced if OFMSW is used as the feedstock. However, analysis for different organics present [such as polycyclic aromatic hydrocarbons (PAHs)] and their toxicity needs to be measured before using it as a fertilizer.

# 3 Summary and Recommendations

MSW can be used as a sustainable energy source after appropriate treatment to increase its high heating value. HTC has shown high potential for energy and resource recovery from biomass and waste materials. It uses the moisture present in wet biodegradable waste for reactions under subcritical conditions. This study focused on different reactions occurring during HTC and options for recovery of energy or value-added products from hydrochar as well as wastewater. Different reactions expected to occur during HTC are hydrolysis, dehydration, decarboxylation, polymerization, and aromatization. The reaction temperature is usually 180–260 °C while the reaction time was also widely varied in previous studies from few minutes to even days. The hydrochar has high heating value similar to brown coal (lignite). Therefore, HTC can be explored on a decentralized scale for wet wastes (>70%) such as, kitchen waste, garden trimmings, sewage sludge, institutional wet waste, and waste from food processing industries. It requires lesser space and time. Moreover, it results in more energy output as compared to biological AD process. The wastewater generated has a high concentration of organics which can be used for biogas generation or resource recovery through bioethanol formation or use as liquid fertilizer. Most of the earlier research in this field had focus on hydrochar properties. However, both the products have not been used for resource recovery simultaneously. The wastewater generated from HTC of OFMSW needs to be analyzed to assess its reusability or resource recovery potential. Furthermore, experimental studies need to be performed to optimize reaction conditions for different types of wastes.

# References

Basso, D., Patuzzi, F., Castello, D., Baratieri, M., Rada, E. C., Weiss-Hortala, E., et al. (2016). Agro-industrial waste to solid biofuel through hydrothermal carbonization. *Waste Management, 47,* 114–121.

Becker, R., Dorgerloh, U., Paulke, E., Mumme, J., & Nehls, I. (2014). Hydrothermal carbonization of biomass: Major organic components of the aqueous phase. *Chemical Engineering and Technology, 37*(3), 511–518.

Berge, N. D., Li, L., Flora, J. R., & Ro, K. S. (2015). Assessing the environmental impact of energy production from hydrochar generated via hydrothermal carbonization of food wastes. *Waste Management, 43,* 203–217.

Berge, N. D., Ro, K. S., Mao, J., Flora, J. R., Chappell, M. A., & Bae, S. (2011). Hydrothermal carbonization of municipal waste streams. *Environmental Science and Technology, 45*(13), 5696–5703.

Funke, A., & Ziegler, F. (2010). Hydrothermal carbonization of biomass: A summary and discussion of chemical mechanisms for process engineering. *Biofuels, Bioproducts and Biorefining, 4*(2), 160–177.

Garg, A. (2014). Mechanical biological treatment for municipal solid waste. *International Journal of Environmental Technology and Management, 17*(2–4), 215–236.

Glaser, B., Haumaier, L., Guggenberger, G., & Zech, W. (2001). The "Terra Preta" phenomenon: A model for sustainable agriculture in the humid tropics. *Naturwissenschaften, 88*(1), 37–41.

Hwang, I. H., Aoyama, H., Matsuto, T., Nakagishi, T., & Matsuo, T. (2012). Recovery of solid fuel from municipal solid waste by hydrothermal treatment using subcritical water. *Waste Management, 32*(3), 410–416.

Kambo, H. S., & Dutta, A. (2015). A comparative review of biochar and hydrochar in terms of production, physico-chemical properties and applications. *Renewable and Sustainable Energy Reviews, 45,* 359–378.

Kim, D., Prawisudha, P., & Yoshikawa, K. (2012). Hydrothermal upgrading of Korean MSW for solid fuel production: Effect of MSW composition. *Journal of Combustion, 2012,* 1–8.

Li, L., Diederick, R., Flora, J. R. V., & Berge, N. D. (2013). Hydrothermal carbonization of food waste and associated packaging materials for energy source generation. *Waste Management, 33*(11), 2478–2492.

Libra, J. A., Ro, K. S., Kammann, C., Funke, A., Berge, N. D., Neubauer, Y., et al. (2011). Hydrothermal carbonization of biomass residuals: A comparative review of the chemistry, processes and applications of wet and dry pyrolysis. *Biofuels, 2*(1), 71–106.

Lin, Y., Ma, X., Peng, X., Yu, Z., Fang, S., Lin, Y., et al. (2016). Combustion, pyrolysis and char $CO_2$-gasification characteristics of hydrothermal carbonization solid fuel from municipal solid wastes. *Fuel, 181,* 905–915.

Lu, L., Namioka, T., & Yoshikawa, K. (2011). Effects of hydrothermal treatment on characteristics and combustion behaviors of municipal solid wastes. *Applied Energy, 88*(11), 3659–3664.

Lu, X., Jordan, B., & Berge, N. D. (2012). Thermal conversion of municipal solid waste via hydrothermal carbonization: Comparison of carbonization products to products from current waste management techniques. *Waste Management, 32*(7), 1353–1365.

Mohapatra, S., Dandapat, S. J., & Thatoi, H. (2017). Physicochemical characterization, modelling and optimization of ultrasono-assisted acid pretreatment of two *Pennisetum* sp. using Taguchi and artificial neural networking for enhanced delignification. *Journal of Environmental Management, 187,* 537–549.

Parshetti, G. K., Chowdhury, S., & Balasubramanian, R. (2014). Hydrothermal conversion of urban food waste to chars for removal of textile dyes from contaminated waters. *Bioresource Technology, 161,* 310–319.

Prawisudha, P., Namioka, T., & Yoshikawa, K. (2012). Coal alternative fuel production from municipal solid wastes employing hydrothermal treatment. *Applied Energy, 90*(1), 298–304.

Reza, M. T., Coronella, C., Holtman, K. M., Franqui-Villanueva, D., & Poulson, S. R. (2016). Hydrothermal carbonization of autoclaved municipal solid waste pulp and anaerobically treated pulp digestate. *ACS Sustainable Chemistry & Engineering, 4*(7), 3649–3658.

Stemann, J., Putschew, A., & Ziegler, F. (2013). Hydrothermal carbonization: Process water characterization and effects of water recirculation. *Bioresource Technology, 143,* 139–146.

Wirth, B., Mumme, J., & Erlach, B. (2012). Anaerobic treatment of waste water derived from hydrothermal carbonization. In *20th European Biomass Conference and Exhibition*, 18–22 June 2012, Milan, Italy.

Yoshikawa, K. (2009). Hydrothermal treatment of municipal solid waste to produce solid fuel. In *7th International Energy Conversion Engineering Conference*, 2–5 August 2009, Denver, Colorado.

Zhao, P., Shen, Y., Ge, S., Chen, Z., & Yoshikawa, K. (2014). Clean solid biofuel production from high moisture content waste biomass employing hydrothermal treatment. *Applied Energy, 131,* 345–367.

# Wastewater Treatment Techniques for Sustainable Aquaculture

Darwin Chatla, P. Padmavathi and Gatreddi Srinu

**Abstract** To support the rapidly growing human population, food production industries such as aquaculture needs horizontal as well as vertical expansion. The rapid growth of global aquaculture industry cannot be overemphasized because environmental and economic limitations hamper this growth. Intensification of aquaculture activities generates excess amounts of organic pollutants that are likely to cause acute toxic effects and long-run environmental risks. Hence, the aquaculture industry has become an axis for criticism from environmental groups because of an apparent negative effect on the environment by the release of wastewater. The routine method of dealing with this problem is the continuous replacement of the pond water through water exchange using clean water. Thus, aquaculture requires not only the supply of clean water but also the release of pollutant-free water for the protection of aquatic environment and reuse of water sources. The main contaminants of wastewater effluent are suspended solids, nitrogenous wastes and phosphates. Therefore, it is obvious that appropriate wastewater treatment processes are needed for sustainable aquaculture development. A number of physical, chemical and biological methods used in conventional wastewater treatment have been applied in aquaculture systems. This review gives an overview about possibilities of treating the wastewater in aquaculture to avoid the pollution and enabling water for reuse.

**Keywords** Aquaculture · Pollution · Wastewater treatment · Environmental risks

## 1 Introduction

Fisheries and aquaculture are the efficient protein production sectors offering ample opportunities to alleviate poverty, hunger and malnutrition (FAO 2017). Globally, India occupies the second position after China with a share of 7.1% world's aquaculture production (FAO 2018). The vast resources in terms of water bodies and

D. Chatla · P. Padmavathi (✉) · G. Srinu
Department of Zoology & Aquaculture, Acharya Nagarjuna University, Nagarjuna Nagar, AP, India
e-mail: padmapin@yahoo.com

© Springer Nature Singapore Pte Ltd. 2020                                         159
S. K. Ghosh (ed.), *Waste Management as Economic Industry Towards Circular Economy*,
https://doi.org/10.1007/978-981-15-1620-7_5_17

species of fish and shellfish in different agro-ecological regions of the country provide for a wide array of culture systems and practices. Inspite of these resources, the rapid growth of aquaculture industry is hampered because of several environmental and economic limitations (Gomez et al. 2018). Extensive method of aquaculture involving the farming of finfish or shellfish in a natural habitat with little inputs and no supplementary feeding has minimum impact on the environment. However, the semi-intensive and intensive culture practices involving the usage of more inputs mostly of high-quality artificial feeds and chemicals (Arvanitoyannis and Kassaveti 2008) lead to the production of large quantities of solid and nutrient wastes into the environment. They are detrimental to the environment by causing eutrophication and toxic to natural aquatic fauna (Wetzel 2001; Kuhn et al. 2010; Cubillo et al. 2016; Srithongouthai and Tada 2017). In addition, aquaculture effluents may be associated with other compounds such as pathogens, heavy metals, hormones or antibiotics, thus generating a risk for human health (Madariaga and Marin 2016; Windi et al. 2016). As a result, the effluents from the aquaculture are now considered as potential pollutants of the aquatic environment and become an axis for criticism from environmental groups. Thus, aquaculture requires not only a continuous supply of clean water but also needs treatment of effluent water before being released into the environment for reuse as well for protecting the biodiversity of natural habitats. Therefore, it is apparent that appropriate wastewater treatment processes are needed for sustainable aquaculture development so as to reduce the negative impact on the environment and provide greater long-term ecological and economic safety for the operation of aquaculture industry. Hence, the present attempt has been made to give an overview of the wastewater treatment techniques in aquaculture for pollution abatement and reuse of water.

## 2   Aquaculture Wastewater

Wastewater generation in aquaculture is a result of its operation from hatcheries and farming systems (Sohail 2003). The types of wastes produced in aquaculture farms are basically similar. However, there are differences in quality and quantity of components depending on the location, species cultured and the culture practices adopted (Antony and Philip 2006). The wastes in hatcheries or aquaculture farms can be categorized as: (1) residual food and fecal matter; (2) metabolic by-products; (3) residues of biocides and biostats; (4) fertilizer-derived wastes; (5) wastes produced during molting and decay of dead organisms and (6) collapsing algal blooms (Sharma and Scheeno 1999). However, the sources of wastewater are primarily from uneaten food and fish feces, which is 30 percent unconsumed dry feed and 30 percent consumed food egested as feces (Axler et al. 1996). The following are the current approaches for improving water quality and wastewater treatment techniques in aquaculture.

# 3   Wastewater Treatment Techniques

## 3.1   Recirculating Aquaculture System (RAS)

Recirculating aquaculture system (RAS) represents a new and advanced technique especially for indoor and laboratory cultures. Instead of traditional method of growing fish outdoors in open ponds and raceways, this system rears fish at high densities in a controlled environment. RAS has been identified as a technology that can achieve high production with minimal ecological impact. Moreover, it offers greater effluent discharge control and also allowing easier recycling of compounds generated in the process (Martins et al. 2010). RAS recycles water by running it through filters to remove fish waste and food, then recirculating it back into the tanks. This saves water and the waste gathered can be used as compost or, in some cases, could even be treated and used on land. In recirculating systems, the daily water exchange rate is reduced to 30 to 50 times when compared to that of an open system (Li et al. 2018). In addition, it can achieve high rates of water reuse by mechanical, biological and chemical filtration and other treatment steps. By this approach, the major toxic pollutants from the culture water can be removed without causing environmental concerns (Gutierrez-Wing and Malone 2006). The renewed interest in recirculating systems is due to their apparent advantages like minimum land and water requirements offering a promising solution to water use conflicts, water quality and waste disposal, and a high degree of environmental control allows year-round growth at optimal rates (Masser et al. 1999). The new water quality technology, testing and monitoring, instrumentation and computer-enhanced system-design programs developed for the wastewater treatment industry have been incorporated and revolutionized our ability for sustainable fish culture (Helfrich and Libey 1991). These concerns will continue to intensify in the future as water demand for a variety of uses escalates. Despite its apparent potential, management of these systems takes high-risk, hence requires expertise. As it involves high operational and maintenance cost, the adoption of RAS among the farming community especially in developing countries is low.

## 3.2   Integrated Multi-trophic Aquaculture (IMTA)

The term multi-trophic refers to the incorporation of species from different trophic or nutritional levels in the same system and is the potential distinction from polyculture systems (Marton 2008). The 'integrated' in IMTA refers to the more intensive cultivation of the different species in proximity of each other, connected by nutrient and energy transfer through water (Turcios and Papenbrock 2014). IMTA is a precision aquaculture in which many options can be developed according to the environmental, biological, physical, chemical, societal and economic conditions. It can be applied to open-water or land-based systems, marine or freshwater systems and temperate or tropical systems (Chopin 2013). The IMTA involves the cultivation of fed species (e.g., finfishes or shrimps fed commercial diets) with extractive species, which utilize the organic (e.g., suspension and filter feeders) and inorganic (e.g., seaweeds or other aquatic vegetation) excess nutrients from fed aquaculture for their growth (Chopin 2013). Here, the by-products or wastes from one species are recycled to become inputs as fertilizers or food for another and to take advantage of synergistic interactions among the species (Chopin et al. 2001, 2008; Troell et al. 2003; Neori et al. 2004; Buck et.al. 2018). IMTA facilitates biomitigation and diversification of fed monoculture practices, by combining them with extractive aquaculture species, to realize the benefits environmentally, economically and socially (Chopin et al. 2012). In addition to create balanced systems for environmental sustainability, it offers a benefit on reducing the water pollution. A working IMTA system can result in greater production based on mutual benefits for the co-cultured species and improved ecosystem health. However, this can be achieved through the appropriate selection and ratios of different species providing different ecosystem functions. Hence, by establishing such integrated cultivation systems, the sustainability of aquaculture may be increased (Chopin et al. 2001; Neori et al. 2004; FAO 2006; Buchholz et al. 2012). Due to its environment-friendly characteristics, the countries like Canada (Ridler et al. 2007; Chopin 2013), China (Fang et al. 2016), the Philippines (Largo et al. 2016), Bangladesh (Kibria and Haque 2018), etc., are using IMTA technology. However, in India, the slow adoption of IMTA technology is in part due to the high initial capital investments and operational costs.

## 3.3   Biofloc Technology (BFT)

Biofloc is a conglomeric aggregation of microbial communities such as phytoplankton, bacteria and living and dead particulate organic matter. Biofloc technology involves the manipulation of C/N ratio to convert toxic nitrogenous wastes into the useful microbial protein and helps in improving water quality under a minimal/zero water exchange system (Ahmad et. al. 2017). By utilizing a microbial community, BFT system acts as an *in situ* biofilter removing dangerous nitrogen compounds from the water (Avnimelech 1999; Crab et al. 2012). Compared to conventional water treatment technologies used in aquaculture, biofloc technology provides a more economical alternative (decrease of water treatment expenses in the order of 30%) system, thus prevents eutrophication and effluent discharge into the surrounding environment. Conventional technologies to manage and remove nitrogen compounds are based on either earthen treatment systems or a combination of solid removal and nitrification reactors (Crab et al. 2007). These methods have the disadvantage of requiring frequent maintenance, and in most instances, the units can achieve only partial water purification. They generate secondary pollution and are often costly (Lezama-Cervantes and Paniagua-Michel 2010). Unlike the conventional techniques such as biofilters, biofloc technology promotes nitrogen uptake by bacterial growth which decreases the ammonium concentration more rapidly than nitrification (Hargreaves 2006; Avnimelech 2009). Moreover, the technology will be useful to ensure biosecurity, as there is no water exchange except sludge removal. The strength of the biofloc technology lies in its 'cradle to cradle' concept as described by McDonough and Braungart (2002), in which the term waste in fact does not exist. Translated in biofloc terms, 'waste' nitrogen generated by uneaten feed and excreta from the cultured organisms is converted into proteinaceous feed available for those same organisms. Instead of 'downcycling,' a phenomenon often found in an attempt to recycle, the technique actually 'upcycles' through closing the nutrient loop. Hence, the water exchange can be decreased without deterioration of water quality, and consequently, the total amount of nutrients discharged into adjacent water bodies may be decreased (Lezama-Cervantes and Paniagua-Michel 2010). Moreover, BFT offers a sustainable aquaculture tool by simultaneously addressing its environmental, social and economic issues. In this context, BFT can also be used in the specific case of maintaining appropriate water quality, and thus, it can reduce the pollution of the pond water. With these positive effects of BFT, it has been widely practicing by shrimp farmers across the world. However, despite its apparent potential, the most obvious disadvantages are continuous aeration of water to maintain high dissolved oxygen of more than 5 ppm, hence high energy cost and need of significantly higher skills and better-equipped laboratories to monitor and operate the biofloc system efficiently. For the efficient organization of this BFT, the ponds must be lined with reinforced concrete or with high density polyethylene (HDPE) sheet. This system can be practiced both in intensive and extensive culture systems.

# 4 Conclusion and Recommendations

The scarcity of water, growing demand for protein food and conflict for land and water usage for the expansion of aquaculture industry are the major problems at global level. To cater to the growing demand for animal protein, intensive aquaculture is one of the major options. But intensification of aquaculture practices will generate a lot of effluents which will damage the aquatic environment. Moreover, the intensification will lead to environmental degradation and socioeconomic conflicts (Ahmad et. al. 2017). In an attempt to minimize the impact of the environmental, health and economic problems associated with aquaculture, wastewater treatment is necessary. Hence, the above-discussed technologies have become increasingly popular as a sustainable alternative for intensification. The requirements for sustainable and eco-friendly aquaculture development can be fulfilled by the use of these technologies. However, no technique is without drawbacks, and also, these techniques are prone to obstacles. A major obstacle is to convince farmers to implement the techniques, since these technologies are expensive ventures particularly when it comes to management. Therefore, the government can educate and support the farmers to implement these technologies in future aquaculture systems for wastewater treatment to reduce the dreadful impact on the environment and to provide greater long-term economic safety for sustainable operation of the industry.

**Acknowledgements** The authors, Darwin Chatla and Gatreddi Srinu are grateful to the UGC for granting BSR (Basic Scientific Research) fellowship. Authors are thankful to the authorities of Acharya Nagarjuna University for providing necessary facilities in completing this work.

# References

Ahmad, I., Rani, A. B., Verma, A. K., & Maqsood, M. (2017). Biofloc technology: An emerging avenue in aquatic animal healthcare and nutrition. *Aquaculture International, 25*(3), 1215–1226.

Antony, S. P., & Philip, R. (2006). Bioremediation in shrimp culture systems. *Naga the World Fish Center Quarterly, 29*(3&4), 62–66.

Arvanitoyannis, I. S., & Kassaveti, A. (2008). Fish industry waste: Treatments, environmental impacts, current and potential uses. *International Journal of Food Science & Technology, 43*(4), 726–745.

Avnimelech, Y. (1999). Carbon/nitrogen ratio as a control element in aquaculture systems. *Aquaculture, 176,* 227–235.

Avnimelech, Y. (2009). *Biofloc technology. A practical guide book* (p. 182). Baton Rouge: The World Aquaculture Society.

Axler, R., Larsen, C., Tikkanen, C., McDonald, M., Yokom, S., & Aas, P. (1996). Water quality issues associated with aquaculture: A case study in mine pit lakes. *Water Environment Research, 68*(6), 995–1011.

Buchholz, C., Krause, G., & Buck, B. H. (2012). Seaweed and man. In Wiencke, C., & Bischof, K. (Eds.) *Seaweed biology: Novel insights into ecophysiology, ecology and utilization* (pp. 471–493). Heidelberg: Springer.

Buck, B. H., Troell, M. F., Krause, G., Angel, D. L., Grote, B., & Chopin, T. (2018). State of the art and challenges for offshore integrated Multi-Trophic Aquaculture (IMTA). *Frontiers in Marine Science, 5,* 165. https://doi.org/10.3389/fmars.2018.00165.

Chopin, T., Buschmann, A., Halling, C., Troell, M., Kautsky, N., Neori, A., et al. (2001). Integrating seaweeds into aquaculture systems: A key towards sustainability. *Journal of Phycology, 37,* 975–986.

Chopin, T., Robinson, S. M. C., Troell, M., Neori, A., Buschmann, A., & Fang, J. G. (2008). Multitrophic integration for sustainable marine aquaculture. *Encyclopedia of Ecology,* 2463–2475. https://doi.org/10.1016/B978-008045405-4.00065-3.

Chopin, T., Cooper, J. A., Reid, G., Cross, S., & Moore, C. (2012). Open-water integrated multi-trophic aquaculture: Environmental biomitigation and economic diversification of fed aquaculture by extractive aquaculture. *Reviews in Aquaculture, 4,* 209–220.

Chopin, T. (2013). Aquaculture, integrated Multi-trophic (IMTA). In Christou, P., Savin, R., Costa-Pierce, B. A., Misztal, I., & Whitelaw, C. B. A. (Eds.) *Sustainable food production.* New York: Springer. https://doi.org/10.1007/978-1-4614-5797-8.

Crab, R., Avnimelech, Y., Defoirdt, T., Bossier, P., & Verstraete, W. (2007). Nitrogen removal techniques in aquaculture for a sustainable production. *Aquaculture, 270*(1–4), 1–14. https://doi.org/10.1016/j.aquaculture.2007.05.006.

Crab, R., Defoirdt, T., Bossier, P., & Verstraete, W. (2012). Biofloc technology in aquaculture: Beneficial effects and future challenges. *Aquaculture, 356,* 351–356.

Cubillo, A. M., Ferreira, J. G., Robinson, S. M. C., Pearce, C. M., Corner, R. A., & Johansen, J. (2016). Role of deposit feeders in integrated multi-trophic aquaculture—A model analysis. *Aquaculture, 453,* 54–66. https://doi.org/10.1016/j.aquaculture.2015.11.031.

Fang, J., Zhang, J., Xiao, T., Huang, D., & Liu, S. (2016). Integrated multi-trophic aquaculture (IMTA) in Sanggou Bay, China. *Aquaculture Environment Interactions, 8,* 201–205.

FAO (Food and Agriculture Organization). (2006). *State of world aquaculture.* FAO Fisheries Technical Paper. No. 500. Rome. http://www.fao.org/3/a-a0699e.pdf.

FAO (Food and Agriculture Organization). (2017). *Aquaculture Newsletter, 56,* 1–64. http://www.fao.org/3/a-i7171e.pdf.

FAO (Food and Agriculture Organization). (2018). *The State of World Fisheries and Aquaculture 2018—Meeting the Sustainable Development Goals.* Rome. http://www.fao.org/3/I9540EN/i9540en.pdf.

Gomez, S., Hurtado, C. F., Orellana, J., Valenzuela Olea, G., & Turner, A. (2018). *Abarenicola pusilla* (Quatrefages, 1866): A novel species for fish waste bioremediation from marine recirculating aquaculture systems. *Aquaculture Research, 49*(3), 1363–1367. https://doi.org/10.1111/are.13562.

Gutierrez-Wing, M. T., & Malone, R. F. (2006). Biological filters in aquaculture: Trends and research directions for freshwater and marine applications. *Aquacultural Engeneering, 34,* 163–171. https://doi.org/10.1016/j.aquaeng.2005.08.003.

Hargreaves, J. A. (2006). Photosynthetic suspended-growth systems in aquaculture. *Aquaculture Engineering, 34,* 344–363.

Helfrich, L. A., & Libey, G. (1991). *Fish farming in Recirculating Aquaculture Systems (RAS).* Virginia State Cooperative Service.

Kibria, A. S. M., & Haque, M. M. (2018). Potentials of integrated multi-trophic aquaculture (IMTA) in freshwater ponds in Bangladesh. *Aquaculture Reports, 11,* 8–16. https://doi.org/10.1016/j.aqrep.2018.05.004.

Kuhn, D. D., Flick, G. J., Jr., Boardman, G. D., & Lawrence, A. L. (2010). *Biofloc: Novel sustainable ingredient for shrimp feed.* Global aquaculture advocate.

Largo, D. B., Diola, A. G., & Marababol, M. S. (2016). Development of an integrated multi-trophic aquaculture (IMTA) system for tropical marine species in southern Cebu, Central Philippines. *Aquaculture Reports, 3,* 67–76.

Lezama-Cervantes, C., & Paniagua-Michel, J. (2010). Effects of constructed microbial mats on water quality and performance of *Litopenaeus vannamei* post-larvae. *Aquacultural Engineering, 42*(2), 75–81. https://doi.org/10.1016/j.aquaeng.2009.12.002.

Li, L., Ren, W., Liu, C., Dong, S., & Zhu, Y. (2018). Comparing trace element concentrations in muscle tissue of marbled eel Anguilla marmorata reared in three different aquaculture systems. *Aquaculture Environment Interactions, 10,* 13–20. https://doi.org/10.3354/aei00250.

Martan, E. (2008). Polyculture of fishes in aquaponics and recirculating aquaculture. *Aquaponics Journal, 48,* 28–33.

Madariaga, S. T., & Marin, S. L. (2016). Sanitary and environmental conditions of aquaculture sludge. *Aquaculture Research, 48,* 1744–1750. https://doi.org/10.1111/are.13011.

Martins, C. I. M., Eding, E. H., Verdegem, M. C., Heinsbroek, L. T., Schneider, O., Blancheton, J. P., et al. (2010). New developments in recirculating aquaculture systems in Europe: A perspective on environmental sustainability. *Aquacultural Engineering, 43*(3), 83–93. https://doi.org/10.1016/j.aquaeng.2010.09.002.

Masser, M. P., Rakocy, J., & Losordo, T. M. (1999). *Recirculating aquaculture tank production systems,* Management of Recirculating Systems. Southern Regional Aquaculture Center. SRAC Publication No. 452.

McDonough, W., & Braungart, M. (2002). *Cradle to cradle: Remaking the way we make things* (p. 193). New York, US: North Point Press.

Neori, A., Chopin, T., Troell, M., Buschmann, A., Kraemer, G. P., Halling, C., et al. (2004). Integrated aquaculture: Rationale, evolution and state of the art emphasizing seaweed biofiltration in modern mariculture. *Aquaculture, 231,* 361–391. https://doi.org/10.1016/j.aquaculture.2003.11.015.

Ridler, N., Wowchuk, M., Robinson, B., Barrington, K., Chopin, T., Robinson, S., et al. (2007). Integrated multi—Trophic aquaculture (IMTA): A potential strategic choice for farmers. *Aquaculture Economics & Management, 11*(1), 99–110.

Sohail, A. S. (2003). Wastewater treatment technology in aquaculture. *World Aquaculture, 34,* 3.

Sharma, R., & Scheeno, T. P. (1999). *Aquaculture wastes and its management.* Fisheries World 22–24.

Srithongouthai, S., & Tada, K. (2017). Impacts of organic waste from a yellowtail cage farm on surface sediment and bottom water in Shido Bay (The Seto Inland Sea, Japan). *Aquaculture, 471,* 140–145. https://doi.org/10.1016/j.aquaculture.2017.01.021.

Troell, M., Halling, C., Neori, A., Chopin, T., Buschmann, A. H., Kautsky, N., et al. (2003). Integrated mariculture: Asking the right questions. *Aquaculture, 226*(1–4), 69–90. https://doi.org/10.1016/S0044-8486(03)00469-1.

Turcios, A. E., & Papenbrock, J. (2014). Sustainable treatment of aquaculture effluents-what can we learn from the past for the future? *Sustainability, 6*(2), 836–856. https://doi.org/10.3390/su6020836.

Wetzel, R. G. (2001). *Limnology: Lake and river ecosystems.* Gulf Professional Publishing.

Windi, I. M., Katariina, P., Timothy, A. J., Christina, L., Karkman, A., Robert, D., et al. (2016). Aquaculture changes the profile of antibiotic resistance and mobile genetic element associated genes in Baltic Sea sediments. *FEMS Microbiology Ecology, 92,* 4. https://doi.org/10.1093/femsec/fiw052.

# Physical, Chemical Analysis of Solid Waste and Energy Recovery from Combustible Waste of Institution—A Case Study of Sri Venkateswara University College of Arts

**Pravallika Kadupula and Munilakshmi Nijagala**

**Abstract** The characteristics of solid waste generated from Sri Venkateswara University College of Arts in Tirupati are presented in this paper. Quantity of solid waste generation, its physical components and chemical characteristics of solid waste and disposal practices are detailed. Further, energy recovery from solid waste of college of arts was found out and higher energy content offers energy recovery options from solid waste generated from the college.

**Keywords** Solid waste · Chemical characteristics · Institutions · Physical components and energy recovery

## 1 Introduction

Municipal solid waste management has become a serious problem because of rapid urbanization and improved economic activities. Increased attention has been given by the government in recent years to handle this problem in a safe and hygienic manner. The improper collection and inappropriate disposal of solid wastes represent a source of water, land and air pollution and pose risks to human health and environment (Tcbonoglous et al. 1993 and Bhide et al. 1983). Therefore, municipal solid waste management (MSWM) is one of the major environmental problems (Lakshmi Narayana Prasad et al. 2009). It involves activities associated with generation, storage, collection, transfer and transport, processing and disposal of solid waste.

---

P. Kadupula · M. Nijagala (✉)
Department of Civil Engineering, Sri Venkateswara University College of Engineering,
Tirupati, India
e-mail: nml.svuce@gmail.com

© Springer Nature Singapore Pte Ltd. 2020
S. K. Ghosh (ed.), *Waste Management as Economic Industry Towards Circular Economy*,
https://doi.org/10.1007/978-981-15-1620-7_5_18

## 2    Details of Study Area

Tirupati, the abode of Lord Sri Venkateswara, is situated at latitude of 130 271 N longitude of 790 E and is spread over an area of approximately 24 km$^2$. The present study area namely Sri Venkateswara University College of Arts is located in S. V. University, Tirupati. The location map of Tirupati is shown in Fig. 1.

In the present study, samples of solid waste are collected from Sri Venkateswara University College of Arts located in Tirupati and per capita solid waste generation was estimated as 0.25 kg/person/day, respectively. Also, solid waste samples were analyzed for certain physical components and chemical characteristics, and the results are presented in Tables 1 and 2. Further, different components in solid waste are subjected to analysis to know their potential for energy recovery (Karthikeyan 2008). From Table 1, it was observed that organic matter content is more than other items. From Table 2, it was observed that the carbon content is more when compared to other parameters.

Mixed solid waste, wood, paper and organic matter are subjected to analysis of energy content by using bomb calorimeter by standard methods to know their calorific value; the same is presented in Table 3.

From Table 3, it was observed that the calorific value of wood and organic matter is higher. It follows from the energy content analysis that the energy recovery from solid waste of Sri Venkateswara University College of Arts is high; above 2000 kcal/kg, therefore, it has a high potential for energy recovery.

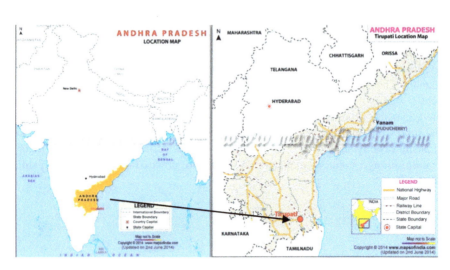

**Fig. 1** Location map of Tirupati

**Table 1** Physical components of solid waste generated from Sri Venkateswara University College of Arts

| Sl. No. | Component | Value (%) |
|---|---|---|
| 1. | Paper | 22.22 |
| 2. | Hair | 0.19 |
| 3. | Plastic | 3.120 |
| 4. | Silt | 8.62 |
| 5. | Aluminum | 0.112 |
| 6. | Glass | 3.3 |
| 7. | Wood | 1.085 |
| 8. | Rubber | 0.138 |
| 9. | Organic matter (dry leaves, food, etc.) | 47.7 |
| 10. | Moisture content | 41 |
| 11. | Density (kg/m$^3$) | 185 |

**Table 2** Chemical characteristics of mixed solid wastes generated from Sri Venkateswara University College of Arts

| Chemical characteristics | Value |
|---|---|
| Carbon (%) | 8.65 |
| Nitrogen (%) | 0.68 |
| Phosphorous (%) | 0.43 |

**Table 3** Calorific values of different types of refuse from Sri Venkateswara University College of Arts

| Component | High calorific value (kcal/kg) |
|---|---|
| Mixed sample | 1708 |
| Wood | 4819 |
| Paper | 3532 |
| Organic matter | 3874 |

# 3   Conclusion and Suggestions

**From the present investigation, following conclusions may be drawn:**

1. Chemical characteristics states that the solid waste of study area contains high carbon content.
2. Energy content analysis states that energy recovery from solid waste of study area is high.

**Some Suggestions from this Study are:**

1. Recycle materials or recover energy from dry wastes.
2. A proper scientific approach should be adopted for collection, transfer and transport, processing and/or disposal of solid waste with material and energy recovery wherever feasible within the campus.

# References

Bhide, A. D., & Sundharshan, B. B. (1983). A text book "Solid waste management" in developing countries.In *Indian National Scientific Documentation Centre*.

Karthikeyan, J. (2008). A study on status of solid waste management in Tirupathi. In *Research project sponsored by government of India and World Bank through TEQIP under services to community and economy*.

Lakshmi Narayana Prasad, P., Karthikeyan, J., & Srivastava, R. C. (2009). SWM and material recovery in an urban area in India—a case study of Tirupathi municipal corporation. In *The 24th International Conference on Solid Waste Technology and Management*, March, Philadelphia, USA.

Tchobanglous, et al. (1993). *Text book "Integrated Solid Waste Management"*. Mc Graw-Hill Publishers, New York.

# Role of Industries in Resource Efficiency and Circular Economy

R. Van Berkel and Z. Fadeeva

**Abstract** Resource Efficiency is concerned with decoupling the growth of human well-being and economic development from the increased use of natural resources and aggravated negative environmental impacts. Resource Efficiency is a linchpin for the transition to Circular Economy, which is ultimately concerned with harmonizing and balancing the cyclic use of materials, both natural and man-made, water and energy in the economy with the long-term carrying capacity of the environment. Resource Efficiency and Circular Economy result from changes in production and consumption systems, and the markets and cultural, normative, policy and regulatory frameworks these operate in. There is compelling evidence of business and economy-wide benefits of greater Resource Efficiency and circularity that can be unlocked by enabling businesses and industries to invest in productivity and innovation, with the three-pronged aim of: maximizing substitution of non-renewable resources; improving the efficiency of use of all natural resources; and perpetually recovering value from all wastes.

**Keywords** Resource Efficiency · Circular Economy · Industry · Clean technology

## 1 Global Imperative

Globally, both the use of natural resources, particularly materials (including chemicals), water and energy, and the discharge of wastes, to land, water and air, have trespassed the finite carrying capacity of the planet (see e.g.: UNEP 2011b; UNEP

The views expressed herein are those of the authors and do not necessarily reflect the views of the United Nations, the United Nations Industrial Development Organization, their respective secretariats or any of their respective member states. Designations such as developed, industrialized, developing and transition are intended for convenience and do not necessarily express a judgement about the stage reached by a particular country or area in its development process.

R. Van Berkel (✉)
United Nations Industrial Development Organization, New Delhi, India
e-mail: r.vanberkel@unido.org

Z. Fadeeva
Office of the United Nations Resident Coordinator, New Delhi, India

© Springer Nature Singapore Pte Ltd. 2020                                      171
S. K. Ghosh (ed.), *Waste Management as Economic Industry Towards Circular Economy*,
https://doi.org/10.1007/978-981-15-1620-7_5_20

2012). Since the turn of the millennium, this is increasingly noticeable at the global and local levels. Weather patterns started to change; water is becoming increasingly scarce; ecosystems are in decline at land and at sea; an increasing number of megacities is choking in air pollution; chemicals are accumulating in food chains; litter is turning remote oceans in plastic soups; etc. Using global hectares as a proxy indicator for the environmental impacts of consumption and production, currently, the world consumes about 1.6 times what the Earth can sustainably provide and absorb in the long run (WWF 2016). It is projected at least two planets Earth are required by 2030 if current trends continue. The link between human well-being and economic development on the one hand and the increased use of natural resources and environmental impacts on the other hand needs to be broken, a notion referred to as 'decoupling'. The International Resource Panel (IRP) highlighted the urgency to combine 'doing more with less resources' (resource decoupling) with 'doing more with less pollution' (impact decoupling) (UNEP 2011a). Decoupling of resource consumption and environmental impact generation is central to both Resource Efficiency and the Circular Economy. As summarized by the IRP (Ekins and Hughes 2016), the term 'Resource Efficiency' is generally used to encompass a number of ideas: the technical efficiency of resource use (measured by the useful energy or material output per unit of energy or material input); the resource productivity or extent to which economic value is added to a given quantity of resources (measured by useful output or value added per unit of resource input); and the extent to which resource extraction or use has negative impacts on the environment (increased resource efficiency implies reducing the environmental pressures that cause such impacts). Resource intensity is the inverse of resource productivity and is therefore measured by resource use per unit of value added. Environmental intensity is similarly the environmental pressure per unit of value added. In its assessment (Ekins and Hughes 2016), the IRP concluded that Resource Efficiency: is essential for meeting the Sustainable Development Goals (SDGs); is indispensable for meeting climate change targets cost-effectively; can contribute to economic growth and job creation; has significant potential affecting key resource flows; and is practically attainable. Earlier (McKinsey Global Institute 2011), the size of the economic opportunity had been estimated at 2.9 trillion USD by 2030, based on practical efficiency options in regard to the use of water, energy, land and steel. Just 15 key opportunities including energy efficiency of buildings, efficient irrigation, tackling food waste and capturing end use steel efficiency, account for 75% of the global economic opportunity. 70–85% of the potential of each is located in developing countries.

A 'Circular Economy' is a systemic approach to economic development designed to benefit businesses, society and the environment. In contrast to the '*take-make-dispose*' linear economy, a Circular Economy is restorative and regenerative by design and aims to decouple growth from the consumption of finite resources. It is based on three principles (Ellen McArthur Foundation 2018):

1. Design out waste and pollution: eliminating the root causes of negative impacts of economic activity, including releases of greenhouse gases and hazardous substances, the pollution of air, land and water, as well as structural waste such as traffic congestion;
2. Keep products and materials in use: to preserve more value in the form of energy, labour and materials, through designing for increased use and utilization, durability, reuse, remanufacturing and recycling to keep products, components and materials circulating in the economy; and
3. Regenerate natural systems: use of renewable resources and their preservation and enhancement, for example, by returning valuable nutrients to the soil to support regeneration or using renewable energy as substitute for fossil fuels.

In its work, the World Business Council for Sustainable Development (WBCSD) emphasizes that the Circular Economy is a new way of looking at the relationships between markets, customers and natural resources (WBCSD 2017). The Circular Economy leverages new business models and disruptive technologies to break and transform the dominant linear economic model. The goal is to retain as much value as possible from resources, products, parts and materials to create a system that allows for long life, optimal reuse, refurbishment, remanufacturing and recycling. (WBCSD 2017) differentiated five business models, respectively: (1) circular supplies (using renewable energy and bio-based or fully recyclable inputs); (2) resource recovery (recover useful resources out of materials, by-products or waste); (3) product life extension (increase life of products through repair, upgrade and resale and through innovation and product design); (4) sharing platform (shared use, access or ownership to increase product use); and (5) product as a service (paid access to product for customers whilst companies retain ownership to increase product use). In a comparable fashion, other authors have proposed different classifications for Circular Economy business models, for example, (Bocken et al. 2016) provide six strategies for slowing, closing and narrowing resource loops, respectively; access and performance model; extending product value; classic long-life model; encourage sufficiency; extending resource value; and industrial symbiosis.

These business strategies combine with technological opportunities provided through rapid and transformational developments in different technology areas. (WBCSD 2017) identified three technology domains with the highest scope for disruptive impact, respectively: digital (to track resources and monitor and improve resource utilization through such technologies as Internet of Things (IoT), big data and blockchain); physical (to utilize and transform resources more efficiently and effectively through such technologies as 3D printing, nanotechnology, advanced materials and energy storage and generation); and biological (to make greater and more efficient use of renewable resources through such technologies as bio-energy, bio-based materials, bio-catalysis, hydroponics and aeroponics).

The Circular Economy builds upon good environment, resource and energy conservation techniques and practices that have already been proven by the United

Nations Industrial Development Organization (UNIDO) and others to benefit developing countries with regard to generating increased income, reducing resource dependency, minimizing waste and reducing environmental footprint (UNIDO 2017a). Transitioning to a Circular Economy is estimated to be able to unlock the global GDP growth of USD4.5 trillion by 2030 and will enhance the resilience of global economies (Lacy and Rutqvist 2015). In dollar terms, the global Circular Economy opportunity (USD4.5 trillion by 2030) represents 37.5% of the estimated total economic opportunity of the SDGs (USD12 trillion by 2030) (BSDC b). In the case of India, the Circular Economy development path could create annual value of USD218 billion in 2030 (equalling 11% of 2015 national GDP) and USD 624 billion in 2050 (equalling 30% of 2015 national GDP), compared with business as usual development scenario, based on assessment of three focus areas: mobility and vehicle manufacturing; food and agriculture; and cities and construction (Ellen McArthur Foundation 2016). This would also allow a significant reduction in the intensity of emissions of greenhouse gases (GHG), 23% by 2030 increasing to 43% by 2050. In dollar terms, Circular Economy opportunity of USD218 billion by 2030 compares to SDG economic opportunity of USD1 trillion in India, i.e. 22%. (BSDC 2017a). Compared with China's current development path, a Circular Economy trajectory could save businesses and households approximately USD 5.1 trillion in 2030 and USD 11.2 trillion in 2040 in spending on high-quality products and services, with regard to the built environment, mobility, nutrition and use of textiles and electronics (Ellen McArthur Foundation 2018). These savings, equivalent to around 14% and 16% of China's projected GDP in 2030 and 2040, respectively, could enable more Chinese urban dwellers to enjoy a middle-class lifestyle. A Circular Economy approach would also reduce the environmental impacts of this lifestyle in China, reflected in large reductions in emissions of greenhouse gases (11% by 2030, 23% by 2040) and in fine particulate matter (10% by 2030, 50% by 2040), and fall in traffic congestion (36% by 2030, 47% by 2040). Moreover, these benefits are enabled by a lower consumption of energy and materials and greater efficiency in the mobility system, which could lessen China's reliance on imported raw materials.

## 2   Industry Opportunity

In operational terms, Resource Efficiency and Circular Economy are essentially twin concepts concerned with improving the creation, preservation and recovery of economic value from natural resources. Resource Efficiency therein stresses the importance of using all natural resources efficiently and prolongedly (to narrow and slow resource cycles) and Circular Economy emphasizes circularity in the use of natural and man-made materials (to further slow and ultimately close resource cycles). As visualized in Fig. 1, taken together, these twin concepts set an industry agenda to: firstly, maximize use of renewable resources; secondly, relentlessly pursue efficiency in the use of all natural resources; and, thirdly, perpetually recover and recycle end of life products and by-products.

**Fig. 1** Industry action
agenda for Resource
Efficiency and Circular
Economy. Adapted from Van
Berkel (2018a) and Nasr
et al. (2018)

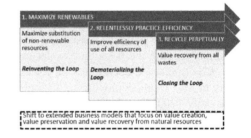

## 3   Maximize Renewables

The lead idea is to ultimately resource man-made systems up to nature's capability
or in other words to produce and consume with renewables harvested or extracted
at rates that are compatible with their long-term availabilities and cycles as well as
maintain balance and resilience of the ecosystems they impact. This can be broadly
achieved by using nature as an input and/or using nature as a mentor.

Nature can be used as an input for supply of materials, energy, water, land/topsoil
and biodiversity. This is well-established practice in many economic sectors, through,
e.g. use of wood and other biomass as source for fuel or fibre, water for energy,
nutrition and sanitation purposes, and topsoil for agriculture and food production.
Many and diverse new applications have come up recently, such as use agro-residues
(e.g. wheat straw) for packaging material or substitute of panel board in furniture
and homewares. There is scope for new applications, for increasing the efficiency
and efficacy of use of renewable resources (with connects to efficiency action of
the industry agenda) and for reducing impacts of extraction. New applications may
include biomaterials, such as bioplastics or bio-solvents, and renewable energy, such
as solar thermal and photovoltaic. Efficiency and efficacy of renewables are enhanced
through advanced use of techniques and practices, such as energy-efficient solar pan-
els and wind turbines, customized high-efficiency bio-fuel furnaces, etc. In terms of
reducing adverse impacts of extraction and harvesting of renewables, maintaining
productivity, diversity and resilience of the source ecosystem is most important. A
growing number of certification schemes are available to enable sustainable sourcing
of renewables, including, for example, Forest Stewardship Council, Marine Stewardship Council and Union for Ethical Biotrade, which are gaining increasing acceptance
and use.

Solar thermal process heat is, for example, promising for many industries that
require process heat at low to medium temperature ranges (up to some 400 °C)
which can be directly achieved through solar thermal collectors (IRENA 2015).
This includes amongst other food processing, dairy, pharmaceutical, chemical and
textiles. Solar process heat was for example demonstrated in the leather sector in
India, both through air heater to dry finished leather achieving 14% reduction in
specific coal consumption with payback of 3.6 years and through water heater for

producing hot process water for leather processing achieving 12% reduction in specific coal consumption with a payback time of 4.5 years (UNIDO b). Solar heating can be scaled up by using lenses and reflectors to concentrate solar radiation, in combination with the tracking of solar movement. Such concentrating solar thermal (CST) units are operational in, for example, dairy sector in India at Mothers Dairy in New Delhi, deploying 16 parabolic dishes of 95 square metres each to produce daily 120,000 L of hot water for the cleaning in place system, and at Amul Dairy in Gandhinagar, using parabolic through collectors with total collector area of 615 square metres to produce steam at 17 kg/cm$^2$ that feeds directly into the steam system (Van Berkel b). At the lower end of the temperature spectrum, S4S Technologies, for example, developed advanced dehydration units that combine conductive, convective and radiative heat transfer for fruit and vegetable drying units (UNIDO 2018).

Using nature as a mentor refers to modelling man-made processes and systems on natural processes, also known as biomimicry (Benyus 1997), and has found its way in a growing number of commercial products, such as self-cleaning surfaces (mimicked from the lotus leaf), efficient rotor blade designs (mimicked from natural vortexes), pigment-free coloured surfaces (mimicked from peacock feathers) and aerodynamic airplane wings (mimicked from bird wings). A recent example refers to the engineering of horizontal gas flames with radiant heat transfer, mimicked from charcoal burning, which has recently been commercialized by Agnisumukh in India for commercial kitchens delivering 30% fuel savings (UNIDO 2018). Watsan engineered a household-level water purification system that does not require energy or chemicals using specifically engineered porous materials that mimicked natural purification materials (UNIDO 2018).

The idea of nature as a mentor also found its way into Green Chemistry and Engineering. Green (or also sustainable) Chemistry and Engineering comprise high-level sustainability strategies for application in the design of product and process chemistries, and of engineering artefacts, particularly industrial plants. Whereas Environmental Chemistry and Engineering deal with minimizing the impacts and risks of pollution and waste on the environment, Green Chemistry and Engineering focus on technological approaches to preventing pollution and reduce the consumption of natural resources, particularly non-renewable resources. Neither Green Chemistry nor Green Engineering are separate sub-disciplines in their own right, yet rather normative, nature inspired frameworks in which sub-disciplinary knowledge, methods and techniques are applied. The application of the twin approaches of Green Chemistry and Engineering has sparked process and product innovations in a range of chemical sectors, as highlighted by for example the US Environmental Protection Agency (USA) under its annual Presidential Green Chemistry Challenge. Recent award-winning industrial applications include (USEPA 2016).

1. Air Carbon$^{TM}$: a biocatalyst system that combines air and methane-based carbon to produce polymers at environmentally friendly ambient conditions, whilst also capturing methane, a potent greenhouse gas. The thermoplastic matches the performance of a wide range of petroleum-based plastics whilst out-competing

on price. Within 24 months of scaling in 2013, AirCarbon™ was adopted by a range of leading companies including Dell, Hewlett-Packard, IKEA, KI, Sprint, The Body Shop and Virgin to make packaging bags, containers, cell phone cases, furniture, and a range of other products;

2. Green Polyurethane™: a safer, plant-based polyurethane for use on floors, furniture and in foam insulation. The technology eliminates the use of isocyanates, which cause skin and breathing problems and workplace asthma. This is already in production and is reducing emissions of volatile organic compounds (VOCs) and costs and is safer for people and the environment;

3. Plantrose® Process: which uses supercritical water to deconstruct biomass provides cost-advantaged cellulosic sugars by using primarily water for conversion reactions. The two-step continuous process deconstructs a range of plant material into renewable feedstocks to produce separate streams of xylose and glucose. After sugar extraction, remaining lignin solids can be burned to supply the bulk of the heat energy required for the process (or utilized in higher-value applications like adhesives or thermoplastics);

4. Faradayic® Trichrome: uses trivalent chromium [Cr(III)], the least toxic and non-carcinogenic form of chromium, in place of hexavalent chromium [Cr(VI)] in the plating baths. The new electrodeposition process alternates between a forward (cathodic) pulse followed by a reverse (anodic) pulse and an off period (relaxation). This allows for thicker coatings from Cr(III) and can be adjusted to affect the structure and properties of the coating. The product exhibits equivalent or improved wear and fatigue performance compared to chrome (VI) coatings. The plating rate can be roughly doubled with about 25% energy use reduction.

Recent cleantech accelerators have uncovered interesting examples displaying elements of Green Chemistry and/or Engineering originating from developing countries. In India, for example, Cellzyme Biotech developed an enzyme biocatalyst system for production of antibiotics at room temperature without using solvents and with higher synthesis yield (UNIDO 2018).

## 4 Relentlessly Practice Efficiency

The guiding idea is to maximize value creation and retention from the resources used in man-made systems, by vigorously pursuing greater efficiency—including through extended duration of use—in the use of all materials, energy and water. Doing so, de facto slows and narrows the resource flows needed to achieve certain functional and economic value.

At enterprise level, this involves Resource Efficient and Cleaner Production (RECP). RECP was introduced to integrate the applications of preventive environmental strategies and total productivity and lean manufacturing methods (UNIDO and UNEP 2010). In strategic terms, RECP is the virtuous process that synergies and realizes progressive improvements in resource (use) efficiency, waste (generation)

**Fig. 2** Resource Efficient
and Cleaner Production as
virtuous process. Image
courtesy of UNIDO

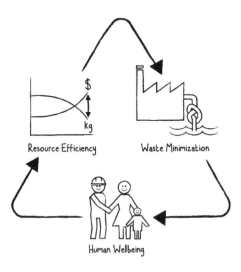

minimization and human well-being. As shown in Fig. 2, these three goals are indeed sequential and mutually synergistic, as higher resource efficiency realizes and facilitates waste minimization, and reduced waste generation, in turn, realizes and enables well-being, and higher well-being, which, in its turn, encourages and enables higher productivity and resource efficiency.

RECP, thus, aims to instil a virtuous and self-propagating synergy amongst resource efficiency, waste minimization and human well-being at enterprise level, and beyond in industrial clusters, regions, value chains and entire production and consumption systems. RECP is operationalized through diverse technical, operational and managerial interventions,that are often loosely grouped into eight categories: good housekeeping; input substitution; better process control; equipment modification; technology change; on-site reuse and recycling; production of useful by-product; and product modification. Table 1 provides for short summary along with indicative water-related examples from textile processing sector.

RECP methods, techniques and practices have a good business case (Van Berkel 2007), arising from reduced expenditures on energy, materials and water, and increased sales from higher productivity and quality, as demonstrated extensively in India (UNIDO 1995) and elsewhere in developing countries over the past 25 years (UNIDO 2015). For example, in Bangladesh, under the Partnership for Cleaner Textiles (PACT), Tarasima Apparels invested US$ 1,398,250 to achieve annual savings of US$655,800 and as a result thereof cut water consumption by 40,560 m$^3$/year, power consumption by 105,000 kWh/year, natural gas consumption by 1,073,280 m$^3$/year, whilst also reducing chemicals consumption and the volume and pollution load of its effluents. The higher cost investments included: installation of thermal oil heaters and of two-ton incineration boiler; using skylights in cutting, washing and finishing units; addition of electro-cascade reactor to improve operation of the effluent treatment plant; and construction of small biogas plant to produce 56 m$^3$ biogas daily from the canteen and other organic wastes (PACT 2015). Rathkerewwa Desiccated

**Table 1** RECP practices illustrated for water use and effluent reduction in textile wet processing (Van Berkel 2017)

| RECP practice | Description | Common water-related example |
|---|---|---|
| Good housekeeping | Maintain a clean, organized and productive ('neat') workplace to eliminate avoidable 'wastage' | • Switch off what is not in use (e.g. taps)<br>• Repair what is broken or leaking (e.g. pipes and hoses)<br>• Remove dry debris before factory wash down |
| Input change | Choose inputs that are efficient, effective and/or pose minimum harm to the environment and health | • Use secondary, recovered water<br>• Use less harmful chemical substances (dyes, detergents, etc.)<br>• Enzyme-enhanced bleaching, scouring |
| Better process control | Monitor and control processes and equipment so that they always run at highest efficiency and with lowest wastage | • Establish and follow standard operating procedures (SOP)<br>• Sub-metre use of water<br>• Install automatic shut-off and overflow prevention valves |
| Equipment modification | Make existing equipment more efficient and less wasteful | • Align and debottleneck production line<br>• Close, hot and cold, process equipment |
| Technology change | Change over to new technology that is more efficient or produces less waste | • Waterless dyeing<br>• Additive, 3D printing |
| On-site reuse and recycling | Use previous 'waste' for similar or alternative purpose in company | • Counter-current or cascaded use of water<br>• Condensate recovery |
| Production of usefull by-product | Convert a previous waste for a useful use elsewhere | • Provide used cooling water for external heating or cooling purposes |
| Product modification | Redesign product to reduce its environmental impact during production, use and/or disposal | • Produce easy care textiles that require minimal water by consumers |

Coconut Industry in Sri Lanka started to use the coconut shells as alternative fuel in its boiler, substituting for the use of coal. Moreover, the company also improved coconut peeling, leading to 50% reduction of kernel losses, halving of water consumption and recovery of coconut oil from wash water. The total investment required was just US$4250 and yielded annual savings of US$315,600 and GHG emission reduction of 900 tons $CO_{2\text{-}eq}$ (SL NCPC 2010). Anning Starch Co in China updated processing equipment in extraction unit (replacement of hammer mill with vertical centrifugal screen by hammer mill with grater and horizontal centrifugal screen) and in refining unit (replacement of disc centrifuge by multistage hydro-cyclone) (CNCPC 2015). Combined with improved integration of the unit operations, implementation of RECP reduced water use by 47%, energy use by 35% and materials use by 11%, which contributed also to reduction of wastewater by 45%. The investment of US$620,000 generated annual savings of US$940,000.

## 5  Recycle Perpetually

The guiding idea is to retain economic value from products at the end of their economic lifetime, which basically serves to embed the man-made economic system into the natural ecosystem. The perpetuality angle is of particular relevance to avoid accumulation in products or dispersion into the environment of potentially hazardous components from increasingly complex materials and products. Value recovery is a multi-pronged agenda to recover value from previously discarded items from every stage of product life cycle, for new product, material, energy or water, with the ultimate aim to retain man-made, non-biodegradable resources in circulation in the economy and release natural, biodegradable resources back to the environment within nature's capacity and at its pace.

Recycling is well-established industrial practice through the 3R methods of reduce, reuse and recycling, for a diversity of materials, such as paper, metals, construction and demolition waste, etc. There is renewed attention to scale up these traditional recycling industries and integrate these in industrial production systems. Large-scale industrial opportunities are for example associated with cement making as this sector can accept and process a diversity of alternative raw materials (e.g. slags, catalysts, etc.) and alternative fuels (including plastics, sludges, tyres, combustible and hazardous waste), in environmentally sound manner, due to the high kiln temperatures, provided best available practices are being deployed (WBCSD and IEA 2013). This though is just one example of the emerging practice of industrial symbiosis which involves the use of one factory's waste as an alternative input material for a nearby factory that can involve waste materials as well as different qualities of wastewater and waste heat. Industrial symbiosis is creating large-scale economic, environmental and operational benefits in industrial complexes in for example Kalundborg (Denmark), Kwinana (Australia), Ulsan (Republic of Korea) and Dalian (China). A further expansion of the symbiosis concept involves the use of municipal waste in industrial processes, leading to industrial and urban symbiosis, as first documented under the Japanese Eco-Town Programme (Van Berkel et al. 2009).

The three Rs also provide a strong framework for cleantech innovation and entrepreneurship. Arvind Textiles in India for example has been able to achieve in its denim production in India 70% circularity in water use (by accepting municipal sewerage as input for its processing plants), 50% circularity in fuels (by using bio and other renewable energy sources) and 20% fibre circularity (post-consumer fibre recovery and reuse), whilst in addition recovering and reusing significant amounts of salts and other processing chemicals. NoWasteTextiles pushed the boundaries of circularity and is already able to produce knitwear from 100% post-consumer recycled garments. Other Indian cleantech start-ups that have wealth from waste as their business model include Saathi (producer of fully biodegradable sanitary pads from waste banana fibre), Aspartika (recovering valuable Omega 3 fatty acids from silkworm pupae) and Brisil (extraction of silica from rice husk ash to produce tyre additive that reduces rolling resistance which in turn saves energy) (UNIDO 2018).

Circularity though extends beyond the traditional environment domain of recycling and resource recovery, into industrial Value-Retention Processes (Nasr et al. 2018). Remanufacturing and comprehensive refurbishment (full-service life Value-Retention Processes) are intensive, standardized industrial processes that provide an opportunity to add value and utility to a product's service life. Repair, refurbishment and arranging direct reuse (Partial Service Life VRPs) are maintenance processes that typically occur outside of industrial facilities and provide an opportunity to extend the product's useful life. IRP assessed VRP for three sectors: automotive components; heavy machinery; and industrial printers (Nasr et al. 2018). It found that at the product-level, remanufacturing and comprehensive refurbishment can contribute to GHG emissions reduction by between 79 and 99%. Similarly, the opportunity for material savings via VRPs is significant: remanufacturing can reduce new material requirement by between 80 and 98%; comprehensive refurbishing saved slightly more materials on average, between 82 and 99%. Repair saved between 94 and 99% and arranging direct reuse largely does not require any inputs of new materials. Cost advantages of VRPs range, conservatively, between 15 and 80% of the cost of new version of the product. An optimized VRP strategy requires that companies adopt new product design processes and priorities. Products must be designed to be durable, upgradable, able to be refurbished or remanufactured and repairable, and these design objectives need to be incorporated early in product planning and business case development stages.

# 6  Outlook

The twin concepts of Resource Efficiency and Circular Economy are reflections of the imperative to decouple economic development from increased use of natural resources and associated emissions, effluents and wastes, from environment, climate, resources and economic angles. There are ample opportunities for producers and consumers to put Resource Efficiency and Circular Economy into practice, through productivity and innovation with the triple aims of maximizing the use of renewable resources, maximizing the efficiency in use of all resources and extending perpetual value recovery and recycling. Whilst praiseworthy results have been achieved in selected enterprises and value chains that are gradually expanding, it is urgent to scale up, speed up and mainstream such good practices to counter today's climate, resources and environmental challenges. This will require transformative change in economic, fiscal, environment, technology and resources policy and practice, and adoption of responsible business practices and sustainable consumption patterns.

# References

Benyus, J. (1997). *Biomimicry: Innovation inspired by nature.* New York: HarperCollins.

Bocken, M., De Pauw, I., Bakker, C., & Van Der Grinten, B. (2016). Product design and business model strategies for a circular economy. *Journal of Industrial and Production Engineering, 33*(5), 308–320.

BSDC. (2017a). *Better business better world: Sustainable business opportunities in India.* London: Business and Sustainable Development Commission.

BSDC. (2017b). *Better business better world: The report of the business and sustainable development commission.* London: Business and Sustainable Development Commission.

CNCPC. (2015). RECP experiences at Anning Starch Co. Beijing: China National Cleaner Production Centre.

Ekins, P., & Hughes, N. (2016). *Resource efficiency: Potential and economic implications; a report for the international resources panel.* Paris: United Nations Environment Programme.

Ellen McArthur Foundation. (2016). *Circular economy in India: Rethinking growth for long term prosperity.* New Delhi: Ellen McArthur Foundation.

Ellen McArthur Foundation. (2018). *The circular economy opportunity for urban and industrial innovation in China.* Beijing: Ellen McArthur Foundation.

IRENA. (2015). *Solar heat for industrial processes: technology brief.* Abu Dhabi: International Renewable Energy Agency.

Lacy, P., & Rutqvist, J. (2015). *Waste to wealth—The circular economy advantage.* New York/London: Palgrave.

McKinsey Global Institute. (2011). *Resource revolution: Meeting the world's energy, materials, food and water needs.* Brussels, San Francisco and Shanghai: McKinsey Global Institute.

Nasr, N., Russell, J., Bringezu, S., Hellweg, S., Hilton, B., Kreiss, C., et al. (2018). *Re-defining value—The manufacturing revolution: remanufacturing, refurbishment, repair and direct reuse in the circular economy; a report of the international resources panel.* Nairobi: United Nations Environment Programme.

PACT. (2015). Cleaner production case study: on the way to sustainability at Tarashimi Apparel Ltd. Dhaka: Partnership for Cleaner Textiles.

SL NCPC. (2010). *Enterprise benefits from resource efficient and cleaner production.* Sri Lanka, Colombo: Sri Lanka National Cleaner Production Centre.

UNEP. (2011a). *Decoupling natural resource use and environmental impacts from economic growth, a report of the working group on decoupling to the international resources panel.* Paris: United Nations Environment Programme.

UNEP. (2011b). *Resource efficiency: Economics and outlook for Asia and the Pacific.* Bangkok: United Nations Environment Programme.

UNEP. (2012). *Global environment outlook 5: Environment for the future we want.* Nairobi: United Nations Environment Programme.

UNIDO. (1995). *From waste to profits*. Vienna: United Nations Industrial Development Organization.

UNIDO. (2015). *National cleaner production centres—20 years of achievement*. Vienna: United Nations Industrial Development Organization.

UNIDO. (2017a). *Circular economy*. Vienna: United Nations Industrial Development Organization.

UNIDO. (2017b). *Reducing pollution load in leather processing—demonstrating cleaner technologies in Kanpur India*. New Delhi: United Nations Industrial Development Organization.

UNIDO. (2018). *A compendium of clean technology innovations in India*. New Delhi: United Nations Industrial Development Organization.

UNIDO and UNEP. (2010). *Taking stock and moving forward: the UNIDO-UNEP national cleaner production centres*. Vienna: United Nations Industrial Development Organization and United Nations Environment Programme.

USEPA. (2016). *Presidential green chemistry challenge winners 1996–2016*. Washington: United States Environmental Protection Agency.

Van Berkel, R. (2007). Cleaner production and eco-efficiency. In D. Marinova, D. Annadale, & J. Phillimore (Eds.), *The international handbook on environmental technology management* (pp. 67–93). Cheltenham: Edgar Elgar Publishing.

Van Berkel, R. (2017). Water efficiency in textile processing; good practices and emerging technologies. India textiles 2017 (p. 4). Ghandinagar: Ministry of Textiles.

Van Berkel, R. (2018a). Materials recycling for circular economy. In *International Conference on Sustainable Growth through Material Recycling—Policy Prescriptions*. New Delhi: National Institution for Transformation of India.

Van Berkel, R. (2018b). Solar heat for industrial processes. Renewable energy invest 2018. Noida: Ministry of New and Renewable Energy.

Van Berkel, R., Fujita, T., Hashimoto, S., & Yong, G. (2009). Industrial and urban symbiosis in Japan: analysis of the Eco-Town Program 1997–2006. *Journal of Environmental Management, 90*(3), 1544–1556.

WBCSD. (2017). *CEO Guide to circular economy*. Geneva: World Business Council for Sustainable Development.

WBCSD and IEA. (2013). *Technology roadmap: Low carbon technology for Indian cement industry*. New Delhi: World Business Council for Sustainable Development and International Energy Agency.

WWF. (2016). *Living planet report 2016: risk and resilience in a new era*. Gland: WWF International.

# Green Temples—Circular Economy in Waste Management

**Mukul Rastogi**

**Abstract** ITC through its social investments brand "Mission Sunehra Kal" initiated sanitation and solid waste management (SWM) interventions, which are aligned to Government of India's Swachh Bharat Mission. ITC's solid waste management (SWM) intervention focuses on "minimizing waste to landfills" and processing waste close to the generator, in a financially sustainable manner. Whilst the company has models for large cities, small towns, rural catchments and temples, operational in 13 districts of 8 states, this paper focuses on its Green Temple SWM model and documents the approach and outcomes of the intervention. Green Temple SWM programme focuses on closed-loop waste management system which processes temple waste into usable products (biogas and compost) that can be used within the temple premises. The first Green Temple SWM intervention was initiated in 2017 in Kapaleeswarar Temple, Chennai, and as part of the project, biogas plant and one bio-composter were installed near the temple premises. The programme focused on capacity building of the temple authorities and local volunteers to take up the programme and independently operate and maintain the initiative in the long run. In addition to this, through Information, Education and Communication (IEC) activity awareness about solid waste management was spread amongst devotees visiting temple, vendors selling temple offerings near the temple premises and local communities to ensure clean and hygienic environment. Post the successful implementation of the Green Temple project in Kapaleeswarar Temple, Chennai, during 2017–18, two new Green Temple projects were initiated—in Anantha Padmanabhaswami Temple, Chennai, and Srirangam Temple, Tiruchirappalli. The Green Temple SWM case study highlights a model that is sustainable, replicable and scalable across locations.

**Keywords** Green Temple · Circular economy · Closed-loop waste management · Zero waste temple and waste to landfill

M. Rastogi (✉)
ITC Limited, Virginia House, Kolkata, India
e-mail: mukul.rastogi@itc.in

© Springer Nature Singapore Pte Ltd. 2020
S. K. Ghosh (ed.), *Waste Management as Economic Industry Towards Circular Economy*,
https://10.1007/978-981-15-1620-7_5_21

185

# 1   Introduction

Waste which is generated through human activity, negatively affects human health, accelerates environmental degradation and adds to existing economic challenges faced by developing countries, including India. In India, waste management initially focused on industrial, commercial and residential waste. However, with time other generators like religious places have also started adding to overall waste generation in the country by virtue of offerings to the deities mainly comprising of coconuts, sweets, flowers, garlands, leaves, millets, clothes, etc. These offerings later in the day litter the temple premises and are thereafter collected and dumped in the closest available unoccupied land or water bodies, leading to pilling up of waste. Such waste pollutes the neighbourhood environment (air, water and land), acts as open breeding ground for mosquitoes and negatively impacts the health of the local communities.

As per census 2011, India has 330 million census houses, 3.01 million places of worship (more than 2 million Hindu temples approximately), and more than the number of schools and colleges (2.1 million). Out of the total waste generated in temples, huge quantity is flowers and leaves which are biodegradable, i.e. more than 80% of the total waste generated is organic, and can be composted (Yadav et al. 2015) to prevent it reaching the landfills.

Traditionally, because of religious beliefs, many avoid throwing flowers and other offerings in the garbage, instead packing it in a plastic bag and then disposing of in water bodies which are even more harmful to the environment. For example, Banaras is known as a holy city in India and it witnesses large inflow of devotees. However, the city has no policy for disposal of tonnes of waste that is generated—nearly 3.5–4 tonnes of waste are generated in the city temples per day (Mishra 2017). A study by Kaushik and Joshi (2011) assessed the waste generated and composition of waste at Mansa Devi and Chandi Devi Temples in the Shivalik Foothills during Kumbh Mela, 2010, and highlighted that total solid waste generated over 7 days was 7615 and 4992.7 kg at Mansa Devi and Chandi Devi Temples, respectively. Further, the composition of waste highlighted that 65% of waste was biodegradable, 12% was non-biodegradable, and 23% was miscellaneous waste.

However, shoots of a revolution in temple waste management are beginning to take shape with temples in Bangalore, Mumbai, etc., and taking steps to compost, ban bringing in of offerings in plastic containers by devotees, etc. and demonstrating varying degree of outcomes in managing waste. This paper attempts to document a workable model and demonstrated outcomes in temple waste management through a circular economy approach.

# 2   Background and Approach

ITC's approach to social investments is premised on a two horizon strategy: Horizon 1 being about creating sustainable livelihoods today and Horizon 2 being about creating a healthy, skilled and educated workforce for tomorrow by focus on improvement in identified Human Development Indices (HDIs). Horizon 2 strategy encompasses interventions in solid waste management amongst others. All interventions aim to mobilize and strengthen community-based organizations (CBOs) for participation and beneficiary contribution for ownership and are implemented in project mode with baseline and impact assessments.

Through its social investment brand "Mission Sunehra Kal", ITC initiated projects on sanitation and solid waste management (SWM) which are aligned to Government of India's Swachh Bharat Mission. The SWM intervention has context-specific models for large cities, small towns, rural catchments and temples, all focused on "minimal waste to landfill" and largely the principle of "treatment close to the generator", so as to reduce transportation and other related costs. ITC's solid waste management programme aims to develop self-sustainable and scalable models, which can be easily adopted across diverse locations.

The Green Temple SWM model of ITC targets creation of "zero waste temples" through the application of appropriate technology and stakeholder involvement and ownership. The project focuses on closed-loop waste management system which processes temple waste into usable products that can be used within the temple premises. The first Green Temple SWM project was initiated during 2016–17 in Kapaleeswarar Temple, Chennai. Post the successful implementation of the Green Temple SWM initiative in Kapaleeswarar Temple, during 2017–18, two new Green Temple projects were initiated—in Anantha Padmanabhaswami Temple, Chennai, and Srirangam Temple, Tiruchirappalli.

# 3 Green Temple SWM Model—The Approach

A baseline assessment of Kapaleeswarar Temple was first commissioned to determine the quantum and composition of waste, current method of disposal and potential technique of waste management. The study findings indicated the presence of a gaushala with cows generating significant quantum of cow dung beside the waste generated from the offerings made by the devotees. Based on the study findings, biogas plant and bio-composter were installed in the temple periphery to deal with the organic waste. Apart from infrastructure support, there was strong focus on sensitizing temple authorities and capacity building of temple workers and the local temple volunteer committees on technological use and maintenance along with sustainable waste management practices.

Some of the key activities undertaken in Green Temple programme are outlined here.

## 3.1   Setting Up Institutional System for Participation and Ownership

Identification of interested people from RWAs, Market Associations and Devotees, organizing them into "Temple Committee", building their capacity on SWM, involving them in planning, implementation of Green Temple activities along with taking waste management awareness to nearby areas and defining their roles for post-project management.

## 3.2   Information, Education and Communication (IEC) Activity to Drive Awareness About the Intervention

Creating larger awareness to all the stakeholders like devotees, schools, street vendors, RWAs, etc., on SWM through various IEC activities, creation and design of comprehensive sustainability plan in partnership with the temple committee and temple authorities.

## 3.3   Technological Support and Operationalization of Infrastructure

Investing in waste management infrastructure like; biogas plant and bio-composter basis the quantum and nature of waste generated, focus on use of biogas in temple kitchen for cooking "Prasadam/Prasad", reduction in number of LPG cylinders on account of biogas and revenue from sale of compost. Tracking and monitoring of the same are to establish proof of concept and outcomes which are replicable and sustainable.

## 3.4   Sustainability, by Handing Over to Temple Authorities

Successful demonstration of technology, ensuring financial sustainability of the model through savings due to reduction in use of LPG cylinders, compost revenue and thereafter, hand-over of Green Temple initiative to temple authorities and the local committee.

To ensure "Plastic and Litter Free Zones", awareness generation regarding solid waste management was provided to vendors selling temple offerings within or near the temple premises with the support of temple committees, with special focus on avoiding use of plastic bags for packaging. In addition to this, awareness and capacity building of local communities and resident welfare associations (RWAs) were undertaken. Children and youth participated in cleanliness drives, sanitation campaigns and other awareness activities including educating street vendors selling offerings, to bring about behaviour change amongst community members.

## 4 Methodology

The paper combines both qualitative and quantitative information to assess the progress of the Green Temple SWM project. The data for the FY 2017–18 is used for assessment purpose.

Pre- and post-intervention waste generated; biogas generated & utilized; compost produced and used; and Cost saving on account of biogas used in temple kitchen for cooking Prasad.

## 5 Results and Outcomes

In Kapaleeswarar Temple, during January 2017, 100% of temple waste was dumped, either in nearby water bodies and/or the corporation landfill, without processing. Post-intervention, during 2017–18, more than 97% of the temple waste is being managed scientifically and used within the temple by converting waste into biogas and compost. During 2017–18, the biogas plant generated around 3102 cum of biogas which was supplied to the temple kitchen for cooking "Prasadam/Prasad" which serves around 3000 devotees and tourists per day. In Kapaleeswarar Temple, installation of biogas plant has on average reduced the monthly LPG cylinder usage by 7 cylinders, equivalent to a monthly average saving of Rs. 8327. The organic waste generated in Kapaleeswarar Temple was composted, and nearly 4.6 tons of compost was produced, which was utilized in the temple Nandavanam (garden) and the balance sold to devotees through the temple committee. The local temple committee and the temple workers are completely equipped with the know-how to manage the temple waste through the process of bio-composting and biogas production. They are aware that besides dealing with waste generated within the temple, a key aspect is reduction in the non-recyclable waste going inside the temple and have been guided to take the initiative to cascade awareness to vendors and devotees to avoid offerings in plastics, etc. Post the successful outcomes of reduction in waste to landfill with accompanying savings in operation costs and readiness of the community organizations to manage the SWM operations, the management of the Kapaleeswarar Temple waste was handed over to the temple authorities and the local volunteer committees in March 2018. The Anantha Padmanabhaswami Temple, Chennai, and Srirangam Temple, Tiruchirappalli, are in early stages of implementation of the model, but similar trends of minimizing waste to landfill (97% and 78%, respectively) are observed from these two temples also. Key results are given in Table 1.

**Table 1** Status of Green Temple initiative

| Green Temple updates for 2017–18 | | | | |
|---|---|---|---|---|
| Details | Unit | Kapaleeswarar Temple, Mylapore | Anantha Padmanabhaswami, Adyar | Sri Ranganatha Temple, Srirangam |
| Month of project launch | | Jan-17 | Oct-17 | Mar-18 |
| Total waste generated in temple | Metric Ton (MT) | 77 | 3 | 9 |
| Total waste composted | MT | 75 | 2.9 | 7 |
| Total waste to landfill | MT | 2 | 0.1 | 2 |
| % of waste processed within temple (overall since inception to March 2018) | % | 97 | 97 | 78 |
| Biogas generated | Cubic metre (Cum) | 3102 | NA | 303 |
| Biogas utilized in temple kitchen | Cum | 2979 | NA | 119 |
| LPG cylinders substituted by temple kitchen | Nos. | 92 | NA | 5 |
| Savings on account of reduction in LPG cylinders | Rs. | 99,918 | NA | 7450 |
| Savings on LPG cylinder per month | Rs. | 8327 | NA | 7450 |
| Compost produced | MT | 4.6 | 0.5 | 0 |

The table gives the outputs for the three Green Temple projects from initiation till March 2018.

Since the waste generated in Anantha Padmanabhaswami Temple comprises of the devotee offering and not of cow dung waste and kitchen waste, only composting intervention is initiated under Green Temple initiative (Table 2).

As ITC's Green Temple programme is designed on the principle of closed-loop circular economy waste management model, more than 90% of waste is converted

**Table 2** Pre- and post-intervention outcomes at Kapaleeswarar Temple, Chennai

| Indicator | Pre-intervention | Post-intervention |
|---|---|---|
| Waste to landfill (%) | 100 | 3 |
| Biogas produced from waste (Cum) | 0 | 3102 cum |
| Average monthly LPG cylinder required in kitchen | 16 | 9 |
| Total savings on account of reduction in LPG cylinders (Rs.) | 0 | 99,918 |
| Compost generated | 0 | 4654 kg |
| Compost used in temple | 0 | 875 kg |
| Compost sold | 0 | 575 kg |
| Total revenue from compost (Rs.) | 0 | 5750 |

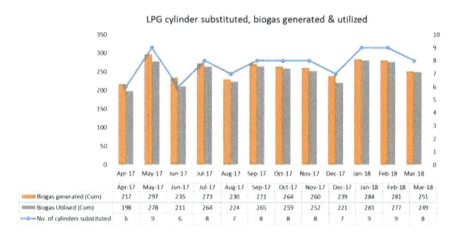

LPG cylinder substituted, biogas generated & utilized

| | Apr-17 | May-17 | Jun-17 | Jul-17 | Aug-17 | Sep-17 | Oct-17 | Nov-17 | Dec-17 | Jan-18 | Feb-18 | Mar-18 |
|---|---|---|---|---|---|---|---|---|---|---|---|---|
| Biogas generated (Cum) | 217 | 297 | 235 | 273 | 230 | 271 | 264 | 260 | 239 | 284 | 281 | 251 |
| Biogas Utilised (Cum) | 198 | 278 | 211 | 264 | 224 | 265 | 259 | 252 | 221 | 281 | 277 | 249 |
| No. of cylinders substituted | 6 | 9 | 6 | 8 | 7 | 8 | 8 | 8 | 7 | 9 | 9 | 8 |

**Fig. 1** LPG substituted, biogas generated and utilized, 2017–18 at Kapaleeswarar Temple

into biogas and compost and is used within the premises of generation. A detail monthly analysis of biogas produced, biogas consumed and saving on number of LPG cylinder usages in temple kitchen is presented in Fig. 1.

# 6 Conclusion

Solid waste management (SWM) in India is a matter of concern for government authorities, and in the near future, temple waste may also become a major menace, along with other wastes. ITC's solid waste management (SWM) programme holistically focuses on technological introduction, embedding behaviour change among its stakeholders and strengthening of community-based organizations (CBOs) so that over time, CBOs can take over and manage the initiative independently in a sustainable manner. ITC's Green Temple SWM model demonstrates self-sustainable and replicable model with the temple acting as a resource centre to impart solid waste management practices to thousands of devotees visiting the temples. The Green Temple SWM model is a circular economy model with focus on reduction and use of the waste generated as a resource, so that minimal waste reaches the landfill. Institutional and financial sustainability, which is a key component of the approach, is clearly demonstrated. Going forward, the Green Temple SWM approach demonstrated at Kapaleeswarar and other two temples has the potential to be replicated sustainably across temples, and such a replication across even a fraction of more than 2 million Hindu temples can have a significant positive impact on reducing the waste challenge in India.

# References

Kaushik, S., & Joshi, D. B. (2011). A comparative study of solid waste generation at Mansa Devi and Chandi Devi Temples in the Shiwalik Foothills, during the Kumbh Mela 2010. Uttarakhand, India.

Mishra, N. (2017). *Temple waste, a concern*. Retrieved from Times of India. http://www.timesofindia.indiatimes.com.

Yadav, I., Juneja, K. S., & Chauhan, S. (2015). Temple waste utilization and management: A review. *International Journal of Engineering Technology Science and Research (IJETSR), 2*(Special Issue), 14–19.

# Faecal Sludge Treatment and Circular Economy: A Case Study Analysis

**Arun Kumar Rayavellore Suryakumar and L. J. Pavithra**

**Abstract** With the success of Swachh Bharat Mission in building toilets for all, faecal sludge is an undeniable outcome of the intervention. The faecal sludge from the on-site sanitation systems such as septic tanks and pits has to be safely managed to convert it to resource-recoverable by-products. From the sanitation circular economy, management of faecal sludge should benefit the environment by generation of less waste, promote innovation and develop markets for reuse of the by-products. The chemical characteristics of faecal sludge show high nitrogen, phosphorus, carbon, moisture as well as high gross calorific value. The paper presents the case study of faecal sludge treatment plant (FSTP) based on thermal technology in context with circular economy. The recovered resources are evaluated in terms of economy and assessments are made to make it a case for economic model through revenues by exploring new avenues for its resource utilization. With more research and development in the field of utilization of the FSTP output resources in diversified areas, FSTPs can play a pivotal role in the sanitation circular economy, especially in addressing the sanitation value chain on the whole. Thermal FSTP utilizes the process thermal energy to achieve biosafety of the by-products, and the design is innovatively integrating nutrient recovery, water recovery and energy recovery during the faecal sludge treatment. The thermal treatment process converts faecal sludge into biochar, treated water and thermal energy. These recovered resources have the potential to positively contribute to the circular economy, which is scientifically analysed in this paper.

**Keywords** Faecal sludge · Circular economy · FSTPs · Resource recovery

## 1 Introduction

The Indian sanitation sector got its acceleration with the launch of Swachh Bharat Mission by the Government of India which aims to achieve an "open-defecation free" (ODF) India by 2 October 2019 by constructing 90 million toilets in rural India. The

A. K. Rayavellore Suryakumar (✉) · L. J. Pavithra
Tide Technocrats, Bengaluru, India
e-mail: rsarunkumar@gmail.com

© Springer Nature Singapore Pte Ltd. 2020
S. K. Ghosh (ed.), *Waste Management as Economic Industry Towards Circular Economy*,
https://10.1007/978-981-15-1620-7_5_22

mission will also contribute to India reaching Sustainable Development Goal 6 (SDG 6), established by the UN in 2015, of securing sanitation for all by 2030. The last 4 years and the next 11 years will witness widespread sanitation systems installations across the country which opens up sanitation circular economy opportunities. The study conducted by Economics of Sanitation Initiative by the World Bank's Water and Sanitation Programme estimated the global benefit at $260 billion in terms of economic gains from sanitation by reducing the burden of health and environmental issues, and even citizens' lost time and productivity. One of the important components of this economy is through the treatment and safe management of the faecal sludge and contributing towards creating self-sustained sanitation businesses and encourage more investment in this sector, reducing its dependency on public and aid funding (WHO/HSE/WSH 2012). The circular economy is the concept in which products, materials (and raw materials) should remain in the economy for as long as possible, and waste should be treated as secondary raw materials that can be recycled to process and reuse (Ghisellini et al. 2016). This is different from linear economy in which waste generated is the last stage of product life cycle. Circular economy promotes sustainable management of materials and energy by reducing the generation of wastes and increases their reuse. The benefits of implementing the concept of circular economy in this field will increase the competition in the market, development of reuse sector and innovations for reusing the wastes, besides fulfilling its main objective of environmental protection.

## 2   Literature Survey

The concept of circular economy in other sectors is followed since years but a thought piece by Toilet Board Coalition made it possible to focus the same principle in sanitation sector. Toilet Board Coalition conducted a feasibility study to find out the potentials of sanitation in the circular economy (Toilet Board Coalition 2016). Faecal sludge being a rich organic material has a lot of potential to drive circular economy, and this has been confirmed by the studies conducted all over the world. Diener et al. evaluated the various resources which can be recovered from faecal sludge treatment to make it a profitable self-sustaining business (Diener et al. 2014). Katukiza et al. reviewed the various technological options available for faecal sludge treatment and assessed their sustainability (Katukiza et al. 2012). Muspratt, et al. assessed the possibility of using faecal sludge as fuel which can be more environmentally and financially rewarding option and recorded as high as 4555 kcal/kg (Muspratt et al. 2014). Energy Alternatives India estimated that the total power potential has thus been estimated to be about 3600 MW from the 0.12 million tons of faecal sludge generated in India (http://www.eai.in/ref/ae/wte/typ/clas/fecal_sludge.html).

# 3 Work

Faecal sludge management is strongly developing into a huge opportunity in coming future. Currently, the paradigm has shifted from treating the waste to primarily on treatment for by-product use and reuse. The by-product uses and reuses mandates biosafety, which is effectively achieved by the thermal process adopted by faecal sludge treatment plant. The FSTP is technically feasible, acceptable to the users, affordable and contribute to health improvement and environmental protection. The FSTP further ensures that the by-products of the process are biosafe for further use and resources derived at the end of the process can be used and contributes towards gaining the self-sustaining entity status.

# 4 Approach and Methodology

FSTP using thermal processes has been establishing into three different cities in India—Narsapur in Andhra Pradesh, Wai in Maharashtra and Warangal in Telangana. The study was carried out at the FSTPs setup by Tide Technocrats, Bengaluru in the above three locations. The quality of septage varies from different locations, and a number of factors influence the septage quality. Hence, 1 L of sample was collected from untreated septage from septage receiving station at regular interval of septage tanker emptying and a composite sample was prepared for analysis. This formed the input to the FSTP, and the characteristics of the input untreated septage influence the quality and type of output or by-products of the FSTP. The FSTP has various components for effective treatment of the faecal waste, which is represented in Fig. 1.

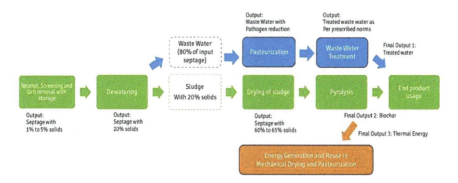

**Fig. 1** Process flow diagram of thermal process for faecal sludge

**Fig. 2** Plant photographs of thermal FSTP

The process consists of six stages, i.e. receipt, screening and storage; solid-liquid separation; treatment of pathogens in the liquid post-separation; treatment of wastewater parameters for meeting PCB norms; moisture reduction in the solids through mechanical drying and treatment of solids and conversion into biochar for biosafety and energy recovery.

The outputs from the FSTP processes are fundamentally

- Treated water, meeting the discharge standards of the Pollution Control Board;
- Biochar, from partial combustion of sludge in the pyrolyser; and
- Thermal energy or heat energy, as a product of pyrolysis.

The detailed chemical analysis of the by-products is carried out at regular intervals, over a period of 6 months. The analysis provided both the purpose of the FSTP—environmental protection and biosafety of the by-products, and thus the spin-off effects resulted in carrying out assessment for effective utilization of the by-products, so as to contribute towards sustained operations.

The overall by-product market assessment is estimated with interactions with sector experts for biochar application, local farmers and other related stakeholders. This assessment forms the primary steps towards establishing resource recovery from faecal sludge management (Fig. 2).

## 5 Significant Results

### 5.1 Chemical Analysis

#### 5.1.1 Septage Characterization

The quality of septage which forms the input to the FSTP varies significantly in different regions of the country and also varies within the city. This depends on various factors, enlisted below:

- Extent of urbanization; space availability
- Water supply—frequency of supply and availability;
- Cultural/traditional practices
- Availability/non-availability of alternate disposal for greywater
- Availability/non-availability of soak-pits post septic tanks

- Desludging intervals
- Size of septic tanks
- Emptying systems in place; water added during desludging the typical character-istics of septage is presented in table.

### 5.1.2   Faecal Sludge Analysis

Post dewatering, the faecal sludge is further processed to prepare for pyrolysis in a mechanical belt dryer. The analysis of the sludge forms the basis for available thermal energy as well as the quantum of fixed carbon in the biochar. The calorific value directly contributes to the energy derivative as well as unburnt carbon.

### 5.1.3   Characteristics of Treated Water

The dewatered filtrate is pasteurized to destruct the helminth eggs and processed in a wastewater treatment system, designed for the appropriate flow rate and output as per the discharge norms of the Pollution Control Board. The treated water, however, has N and P, which can be used for any agri-related activities.

### 5.1.4   Characteristics of Biochar

Biochar samples were sent for analysis as per SWM Rules, 2016 (India), to confirm its suitability for use in agriculture and related field applications. The results are given in Table 4.

## 6   Thermal Energy

The pyrolyzer generates approximately 91 kW/h from the partial combustion of the faecal sludge. Out of 91, 59 kW/h is used collectively utilized by pasteurization for making the treated water biosafe, and in the dryer unit. Pasteurization unit uses 25 kW and dryer unit uses 34 kW respectively, of the available thermal energy as shown in Fig. 3.

## 7   Quantity of Resources Generated

A typical 15,000 l of septage treatment facility with TS in the range 3–4% generates following quantities of by-products as given in table.

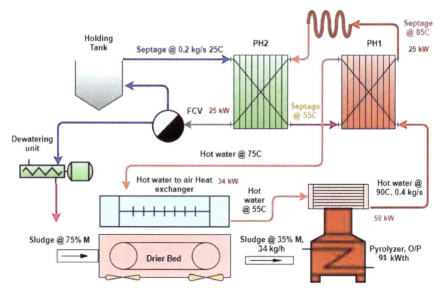

**Fig. 3** Energy balance diagram of thermal FSTP

# 8 Analysis and Discussions

Results of septage analysis (Table 1) show high COD and BOD values of septage which range from 112,000 mg/l to 68,450 mg/l, respectively. The total nitrogen values were reported as 210 and ammonical nitrogen was reported as 210 mg/l. Total phosphorous values were reported 22.1 mg/l. These values agree to the vales reported by Kone et al. who conducted the study of faecal sludge characterization for developing countries (Koné and Nelson 2010). Septage properties are variable

**Table 1** Septage characterization

| Parameter | Unit | Average values |
|---|---|---|
| pH | – | 8.67 |
| BOD | mg/L | 68,450 |
| COD | mg/L | 112,000 |
| TS | % | 3.46 |
| TSS | mg/l | 8.36 |
| TVS | mg/l | 7100 |
| Turbidity | NTU | >800 |
| $NH_3$–N | mg/l | 210 |
| N total | mg/l | 300 |
| Total | mg/L | 22.1 |
| Faecal coliform | MPN/100 mL | $16 \times 10^6$ |

and location dependent due to different structural and environmental conditions and personal hygiene habits, traditions, water availability, etc. Table 2 shows the analysis results of faecal sludge which shows high calorific value of faecal sludge which was estimated an average 3258 kcal/kg. Ash, fixed carbon and volatile solids were estimated 28.43%, 6.63% and 64.09%, respectively. Ultimate analysis of C, H, O, N, S average values was reported 37.32, 5.78, 15.24, 2.51 and 0.832%, respectively. Same study was conducted by Barani et al. also, who also reported similar values in southern part of India (Gopinath et al. 2013). Table 3 shows the results of treated water which are compared to the effluent discharge standards of Ministry of Environment, Forest and Climate change notification, 2017. As per discharge standards, the BOD, TSS and pH values of treated water should be lesser than 30, 100 and 6.5–9.0 while treated water results showed the values 18, 12 and 6.96, respectively. This water can be reused for agriculture and land irrigation, industrial purposes, toilet flushing and groundwater replenishing. The use of treated wastewater for irrigation purpose in agriculture can replace of the agriculture demand and reduction of local water stress. In addition, nutrients contained in the treated water reduce the need for application of commercial fertilizers. Treated water can be used for urban water reuse for non-potable uses such as residential irrigation, fire protection, car wash and toilet flushing. This water can be used for directly recharging groundwater or surface water sources also.

**Table 2** Faecal sludge analysis

| Parameter | Sample 1 | Sample 2 | Sample 3 | Sample 4 | Sample 5 | Average | Max. | Min. |
|---|---|---|---|---|---|---|---|---|
| Calorific value (kcal/kg) | 3290 | 2460 | 5260.6 | 2320 | 2960 | 3258.12 | 5260.6 | 2320 |
| Ash (%) | 30.08 | 29.28 | 21.95 | 30.21 | 30.64 | 28.43 | 30.64 | 21.95 |
| Fixed carbon (%) | 3.072 | 8.96 | 1.13 | 7.63 | 12.37 | 6.63 | 12.37 | 1.13 |
| Volatile matter (%) | 69.72 | 60.92 | 78.05 | 61.38 | 50.41 | 64.09 | 78.05 | 50.41 |
| Carbon (%) | 27.47 | 43.29 | 30.86 | 41.19 | 43.8 | 37.32 | 43.8 | 27.47 |
| Oxygen (%) | NA | 14.39 | NA | 14.21 | 17.12 | 15.24 | 17.12 | 14.21 |
| Hydrogen (%) | 4.53 | 6.86 | 4.69 | 6.72 | 6.1 | 5.78 | 6.86 | 4.53 |
| Nitrogen (%) | 3.33 | 1.56 | 4.53 | 1.63 | 1.5 | 2.51 | 4.53 | 1.5 |
| Sulphur (%) | 1.18 | 0.42 | 1.23 | 0.39 | 0.94 | 0.832 | 1.23 | 0.39 |

**Table 3** Treated water analysis

| Parameter | Typical | [a]Standards |
|---|---|---|
| pH | 6.96 | 6.5–9.0 |
| Colour (1:100 ratio) Hazen | 40 | – |
| BOD (mg/l) | 18 | 30 |
| COD (mg/l) | 90 | – |
| TS (mg/l) | 0.124 | – |
| TSS (mg/l) | 12 | <100 |
| Turbidity (mg/l) | 4 | – |

[a]Ministry of Environment, Forest and Climate change notification, 2017

Table 4 shows the analysis results given for biochar. The analysis results of biochar are compared with Solid Waste Management Rules, 2016. Many studies show that biochar as a soil amendment gives better results rather than directly using the crop residues. It also has a long life in soil and is more effective in sequestering carbon besides improving other soil properties like water holding capacity and nutrient availability (Gopinath et al. 2013). Usage of biochar as soil amendment depends upon its method of application, rate and soil quality. In some parts of India, cases

**Table 4** Biochar analysis in comparison with SWM rules, 2016 (India) organic compost standards

| Sl. No. | Parameters | Units | Biochar | SWM rules, 2016 |
|---|---|---|---|---|
| 1. | pH | Of 5% suspension | 7.5 | 6.5–7.5 |
| 2. | Colour | – | Black | Dark brown to black |
| 3. | Odour | – | No | Absence of foul odour |
| 4. | Moisture | % | 7 | 15–25 |
| 5. | Potassium as K | % | – | Minimum 0.4 |
| 6. | Nitrogen as N | % | 1.96 | Minimum 0.8 |
| 7. | Phosphorous as P | % | 0.002 | Minimum 0.4 |
| 8. | Lead as Pb | mg/kg | 0.029 | 100 |
| 9. | Zinc as Zn | mg/kg | 0.19 | 1000 |
| 10. | Cadmium as Cd | mg/kg | <0.001 | 5 |
| 11. | Copper as Cu | mg/kg | 0.007 | 300 |
| 12. | Nickel as Ni | mg/kg | 0.007 | 50 |
| 13. | Total chromium as Cr | mg/kg | 0.041 | 50 |
| 14. | Particle passing through 4 mm | % | 100 | Minimum 90% material should pass through 4.0 mm IS sieve |
| 15. | Bulk density | g/cm$^3$ | 1.76 | <1 |
| 16. | Lead as Pb | mg/kg | 0.029 | 100 |
| 17. | Mercury | mg/kg | 0.008 | 0.15 |

**Table 5** Estimated resource output of FSTP per day

| Sl. No. | Resource | Quantity/day |
|---------|----------|--------------|
| 1. | Biochar | 45 kgs |
| 2. | Treated water | 13 cum. |
| 3. | Heat energy | 590 kW/day |

have been noted where tribals living in the parts Odisha, Jharkhand and West Bengal states, in India, use of biochar in increasing the crop production. They mix charcoal with farmyard manure (pellets of small ruminants/cattle dung) and add to the red lateritic soils which are then, else less fertile. They cultivate vegetables and green salad in the well-fenced plots of about 1 acre in size. In Indian conditions, there is an immense scope for converting millions of tonnes of faecal sludge generated into biochar and use the same for enriching soil carbon. According to EAI estimates, about 0.12 million tons of faecal sludge is generated in India per day. Efficient and sustainable treatment of faecal sludge is main concern in rural and urban India. Mostly faecal sludge and septage are discharged either in nearby waterbodies or disposed in agricultural fields, which degrade the environment or pose biosafety challenges. If this faecal sludge is converted to biochar, about 0.13 million tons of biochar can be produced annually. During the production of biochar from faecal sludge, all the pathogens are getting destructed and make it biosafe to use in agricultural fields. The important quality of biochar as a soil amendment is its highly porous structure, improved water retention capacity and increased soil surface area. It is important to note that there is a wide variety of char products produced industrially.

The chemical characterization of raw materials and output resources indicates the suitability of by-products to the agriculture. Table 5 shows the estimated resource output of FSTP which is calculated based on input septage TS 2.8% and accordingly revenue generation from FSTP was calculated. The prices of the resources were decided based on current market prices.

# 9 Economic Evaluation of FSTP Output Resources

Interactions were held with the concerned stakeholders such as farmers, professors of Agriculture University and solid waste management practitioners to assess the utility value of biochar and treated water in the agriculture sector. It is widely acknowledged that 1 acre of irrigated land requires about 10,000–12,000 l of water every day, which clearly matched the output from the FSTP, in terms of the quantity. The quality was perceived better than water, because of the presence of N and P.

Similarly, for biochar, the broader consensus was to mix the fixed carbon-rich biochar with city compost from SWM processing and this enhances the quality of city compost. A mixture of 1 part of biochar to 9 parts of city compost is the recommended mix. Thermal energy is equivalent to electrical energy in its value, and hence its direct co-related electricity rate is adopted for the quantification of the value of the output resources from the FSTP. For the quantification purposes, the leftover thermal energy after the process utilization is only considered. Accordingly, the following tank provides the value of the overall recoverable. The assumption is a 300-day operation of the FSTP.

| Sl. No. | Use | Price | Income/month | |
|---------|-----|-------|--------------|--|
| 1. | Biochar | 5/kg | Rs. 1500 | Rs. 67,500 |
| 2. | Treated water | 5/KLD | Rs. 1500 | Rs. 67,500 |
| 3. | Heat energy | 5/kW | Rs. 25,000 | Rs, 300,000 |
| | Total | | Rs. 27,000 | Rs. 435,000 |

## 10  Conclusion/Recommendations

Clearly, the monthly income of Rs. 27,000 would not be able to sustain the operations cost of the FSTP, which is estimated to be about Rs. 100,000–Rs. 120,000 per month, including manpower and consumables. Evidently, the direct values of the recoverable resources would not suffice the balance in the sanitation circular economy to create sustainable operations and encourage private interest in the sector. This definitely provides necessary impetus for innovative thinking for utilization of the by-product resources and enhances the output value. Biochar, for instance, can be utilized for a variety of purposes. Its high surface area of fixed carbon makes it a good material for filters. Biochar is under investigation as an approach to carbon sequestration, as it has the potential to help mitigate climate change. These alternate uses of biochar can enhance the value, thus providing better returns to the same quantity.

Similarly, treated water when used in a greenhouse and for commercial crops like ornamental flowers, medicinal plants, etc., the product value will reflect to the treated water value, and thus create a parallel economy. This is the fundamental principle of circular economy, keeping the outputs in the economy system for a longer duration, thereby creating multiple utilization of by-products. From the sanitation for all perspective, clearly FSTPs are a key element in completion of the sanitation value chain, which provides end-to-end solution of septage treatment. The resources produced by the FSTPs are also useful components as evident from its chemical characterization. The FSTPs can play an important role in helping cities towards a sustainable future, characterized by circular flow of water, waste, material and energy. Present case study shows that the economic output of the resources is very less as compared to capital cost and operation and maintenance cost, which cannot make FSTPs a self-sustained entity. Though the economic gains from having a FSTP is apparent from the reduced burden on human health and the environment, the circular economy context with the value addition from the treated by-products will ensure sustenance of the operations of the plant to a certain extent.

# References

Diener, S., Semiyaga, S., Niwagaba, C. B., Muspratt, A. M., Gning, J. B., Mbéguéré, M., et al. (2014). A value proposition: Resource recovery from faecal sludge—Can it be the driver for improved sanitation? *Resources, Conservation and Recycling, 88,* 32–38.

Ghisellini, Patrizia, Cialani, Catia, & Ulgiati, Sergio. (2016). A review on circular economy: the expected transition to a balanced interplay of environmental and economic systems. *Journal of Cleaner Production, 114,* 11–32.

http://www.eai.in/ref/ae/wte/typ/clas/fecal_sludge.html.

Katukiza, A. Y., Ronteltap, M., Niwagaba, C. B., Foppen, J. W. A., Kansiime, F., & Lens, P. N. L. (2012). Sustainable sanitation technology options for urban slums. *Biotechnology Advances, 30*(5), 964–978.

Koné, D., Cofie O. O., & Nelson, K. (2010). *Low-cost options for pathogen reduction and nutrient recovery.* Faecal Sludge.

Murray Muspratt, A., Nakato, T., Niwagaba, C., Dione, H., Kang, J., Stupin, L., et al. (2014). Fuel potential of faecal sludge: calorific value results from Uganda, Ghana and Senegal. *Journal of Water, Sanitation and Hygiene for Development, 4*(2), 223–230.

Srinivasarao, Ch., Gopinath, K. A., Venkatesh, G., Dubey, A. K., Wakudkar, H., Purakayastha, T. J., et al. (2013). Use of biochar for soil health enhancement and greenhouse gas mitigation in India: Potential and constraints. Central Research Institute for Dryland Agriculture—National Initiative on Climate Resilient Agriculture.

Toilet Board Coalition. (2016). *Sanitation in the circular economy—Transformation to a commercially valuable, self-sustaining, biological system,* A thought piece from the Toilet Board Coalition, November 2016.

WHO/HSE/WSH (2012). Global costs and benefits of drinking-water supply and sanitation interventions to reach the MDG target and universal coverage.

CPSIA information can be obtained
at www.ICGtesting.com
Printed in the USA
LVHW081039200420
654119LV00002B/39

9 789811 516191